ISO 9001:2015 Explained

Also available from ASQ Quality Press:

How to Audit the Process Based QMS, Second Edition
Dennis R. Arter, Charles A. Cianfrani, and John E. "Jack" West

ISO 9001:2008 Explained and Expanded
Charles A. Cianfrani and John E. (Jack) West

Unlocking the Power of Your QMS: Keys to Business Performance Improvement
John E. (Jack) West and Charles A. Cianfrani

ISO 9001:2015 Internal Audits Made Easy, Fourth Edition
Ann W. Phillips

ISO Lesson Guide 2015, Fourth Edition
J.P. Russell and Dennis R. Arter

How to Establish a Document Control System for Compliance with ISO 9001:2015, ISO 13485:2016, and FDA Requirements
Stephanie L. Skipper

ISO 9001:2015 for Small and Medium Sized Businesses, Third Edition
Denise Robitaille

Process Driven Comprehensive Auditing, Third Edition: A New Way to Conduct ISO 9001:2015 Internal Audits, Third Edition
Paul C. Palmes

Using ISO 9001 in Healthcare, Second Edition: Applications for Quality Systems, Performance Improvement, Clinical Integration, Accreditation, and Patient Safety
James M. Levett, MD and Robert G. Burney, MD

The Quality Toolbox, Second Edition
Nancy R. Tague

The Certified Manager of Quality/Organizational Excellence Handbook, Fourth Edition
Russell T. Westcott, editor

The ASQ Auditing Handbook, Fourth Edition
J.P. Russell, editor

To request a complimentary catalog of ASQ Quality Press publications, call 800-248-1946, or visit our website at http://www.asq.org/quality-press.

ISO 9001:2015 Explained

Fourth Edition

Charles A. Cianfrani
John E. (Jack) West

ASQ Quality Press
Milwaukee, Wisconsin

American Society for Quality, Quality Press, Milwaukee 53203
© 2015 by ASQ
All rights reserved.
Printed in the United States of America
20 19 18 17 16   5 4 3

Library of Congress Cataloging-in-Publication Data

Cianfrani, Charles A., author.
  ISO 9001:2015 explained / Charles A. Cianfrani, John E. (Jack) West.
  pages cm
  Includes index.
  ISBN 978-0-87389-901-7 (alk. paper)
  1. ISO 9001 Standard. 2. Quality control—Standards. I. West, Jack,
  1944– author. II. Title.
  TS156.6.C45 2016
  658.5′620218—dc23
                                                        2015033095

Publisher: Lynelle Korte
Acquisitions Editor: Matt Meinholz
Managing Editor: Paul Daniel O'Mara
Production Administrator: Randall Benson

ASQ Mission: The American Society for Quality advances individual, organizational, and community excellence worldwide through learning, quality improvement, and knowledge exchange.

Attention Bookstores, Wholesalers, Schools, and Corporations: ASQ Quality Press books, video, audio, and software are available at quantity discounts with bulk purchases for business, educational, or instructional use. For information, please contact ASQ Quality Press at 800-248-1946, or write to ASQ Quality Press, P.O. Box 3005, Milwaukee, WI 53201–3005.

To place orders or to request a free copy of the ASQ Quality Press Publications Catalog, visit our website at http://www.asq.org/quality-press.

 Printed on acid-free paper

Quality Press
600 N. Plankinton Ave.
Milwaukee, WI 53203-2914
E-mail: authors@asq.org

ASQ The Global Voice of Quality®

Contents

List of Figures and Tables

Preface

The ISO 9000 series of quality management system (QMS) standards was initially issued in 1987 and reissued with minor revisions in 1994. A major revision was issued in 2000 to update the standards and to make the documents more user-friendly. ISO 9001:2008 was the fourth edition of ISO 9001.

The 2015 edition of ISO 9001 has been modernized to update terminology and content to meet current and anticipated user needs. The major emphasis of ISO 9001:2015 is still consistent provision of products and services that meet customer and applicable statutory and regulatory requirements. The latest edition of ISO 9001 incorporates directly or indirectly all the requirements of its predecessor but the language may appear peculiar or unfamiliar. It also contains requirements that may be new for many organizations. The old, the new, and the repackaged requirements are addressed in this book.

A few areas that should receive your particular attention as you address the requirements of ISO 9001:2015 include:

- Understanding the needs of the customer including consideration of internal and external issues and the needs and expectations of interested parties beyond a direct customer

- Considering processes related to risks and opportunities

- Understanding needs, meeting requirements, and monitoring information related to customer satisfaction

- Use of a process approach, which considers all work in terms of suppliers, inputs, processing activities, process interactions, resources, outputs, and customers

- Managing a system of effective processes rather than documenting the system in procedures

- Emphasizing the role of top management

- Setting measurable objectives and measuring product and process performance

- Analyzing and using data to define opportunities for improvement

- Continual improvement of processes and of QMS effectiveness

- Use of wording that can be understood in all product sectors—not just hardware

- New terminology especially for documented information

- The lack of specific requirements for documented information in several places

- The appearance of the elimination of preventive action (which has been replaced by the introduction of risk-based thinking)

All of the above are addressed in detail in this book. As was our objective in the previous three editions, we strive to provide a context for all requirements to enable you to develop and deploy processes that will strengthen your QMS. Getting or retaining a certificate is not the real objective. Satisfied customers and organizational sustainability should be primary objectives for the organization.

This book addresses the needs of:

- Users and organizations seeking a general understanding of the contents of ISO 9001:2015

- Users and organizations desiring guidance to ensure their ISO 9001:2015 QMS meets the new version requirements

- Users and organizations considering the use of ISO 9001:2015 as a foundation for the development of a comprehensive QMS

- Educators who require a textbook to accompany a training class or course on ISO 9001:2015

- Auditors who desire to increase their level of auditing competence

This book explains the meaning and intent of the requirements of ISO 9001:2015 and discusses the requirements as they relate to each of the product categories. Where appropriate, it includes an elaboration of why the requirements are important. It also includes typical audit-type questions that an organization may consider to assess conformity to internal needs and ISO 9001 requirements.

Recommendations for implementation are also included. Each clause has a section on tips for implementation. New tips have been added as appropriate to accommodate new terminology and requirements. ISO 9001:2015 provides requirements for what activities and processes are needed; it does not tell the user how to carry out these requirements. Therefore, the authors have included some recommendations for implementation actions that have proven to be successful. It should be recognized that this implementation guidance may go beyond the requirements of ISO 9001:2015. It is provided because much of what is needed for a successful QMS involves how the requirements are implemented, not the requirements themselves.

There is also a new chapter, "Integrating the Process Approach and Systems Thinking."

The contents of ISO 9001:2015 have been paraphrased in this book. Paraphrased text by its very nature can introduce differences in understanding and interpretation. This book should be used in conjunction with:

- ISO 9001:2015 *Quality management systems— Requirements*

- ISO 9000:2015 *Quality management systems— Fundamentals and vocabulary*

This book is intended to facilitate an understanding of the process approach and related systems thinking and how to apply it in any organization to create, deploy, and improve a QMS. The intended outcome should be a QMS that achieves internal operating effectiveness and improved performance as viewed by customers and other interested parties.

For simplicity, the book often refers to specific editions of ISO 9001 by the year only (i.e., ISO 9001:2015 is referred to as 2015).

1

ISO 9001:2015: What's New and How to Get Started on the ISO 9001:2015 Journey

Plus ça change, plus c'est la même chose. (The more things change, the more they stay the same.)

—French Proverb

Ah, is it just me or does anybody see
The new improved tomorrow isn't what it used to be
Yesterday keeps comin' 'round, it's just reality
It's the same damn song with a different melody

—Jon Bon Jovi

BACKGROUND

Before addressing ISO 9001:2015, the fifth edition of ISO 9001, let us consider the events that brought us from the first edition, in 1987, to the current time.

In 1987 the first edition of ISO 9001 was published by the International Organization for Standardization (ISO). That standard created a common worldwide language for quality assurance. Two additional International Standards were published at the same time that were derivatives of, and less comprehensive than, ISO 9001. The intent at that time was to provide a global set of minimum requirements for quality systems. If an organization could demonstrate conformity with this minimum set of requirements, its customers anywhere in the world could have confidence in the products and services that the organization provided, and international trade would be facilitated.

The first edition did not include the words "quality assurance" or "quality management" in its title—just "quality systems." Over the years the standard has changed in many ways in both its form and content, but its essence and intent have remained the same—the articulation of a minimum set of requirements that an organization should consider in structuring and deploying its quality management system (QMS). The second edition of ISO 9001 was published in 1994, the third edition in 2000, and the fourth in 2008.

In its initial release, ISO 9001 was created as one element of a family of standards—ISO 9000, ISO 9001, and ISO 9004. ISO 9001 was a requirements document, meaning that it stated requirements. ISO 9000 and ISO 9004 were guidance documents that did not contain requirements. The primary use of ISO 9000 was as a quality dictionary, although it contained much additional quality-related information. ISO 9004 was a guideline document that provided its readers with areas and processes an organization could consider to expand the breadth and depth of its quality system. ISO 9004 has always been intended to serve as a stepping-stone for organizations that desired to expand their quality systems beyond a minimum set of processes to world-class levels of quality.

We now have the fifth edition of ISO 9001 available to us.

WHAT IS NEW (IF ANYTHING) IN ISO 9001:2015?

The question many of us ask about ISO 9001:2015 is "Are there any new requirements in ISO 9001:2015?" The answer to this question is "It depends." If an organization has a QMS designed and deployed to meet or exceed the intent of the requirements contained in ISO 9001:2008, the answer to the question could be "Although several new terms appear in 2015, and the structure and many of the words appear to be different, in many cases these new words do not provide new requirements in the 2015 edition." This is the belief of the authors regarding the content of the latest edition. If, however, an organization has achieved only minimal conformity with the requirements of ISO 9001:2008 (e.g., just to pass an outside certification audit), the answer would be quite different and along the lines of "We will need to develop and deploy several new processes in order to be able to provide objective evidence of compliance." For many organizations, the actual situation will be somewhere between these two extremes.

In Chapters 2–9 we provide details of the content of each clause of ISO 9001:2015. For each clause we provide you with the following information to enhance your understanding of the explicit and implicit requirements and to facilitate your consideration of the options you have to develop and deploy compliant value-adding processes:

- What does the clause say? (paraphrased)
- What does the clause mean?
- Consider the potential interactions as they apply to your QMS
- Implementation tips

- Questions to ask to assess conformity

- Definitions (refer to ISO 9000:2015)

- Considerations for documented information to be maintained and/or retained

In addition, after addressing each of the requirements clauses, we provide chapters on the following topics related to the requirements:

- Annexes A and B of ISO 9001:2015

- Auditing implications (internal and external)

- Integrating the process approach and systems thinking

- Selected sector applications of ISO 9001:2015 (automotive, aerospace, and telecommunications)

To start our journey to obtain a working knowledge of ISO 9001:2015, we provide an overview of the content that we believe may appear to be new to many users of the standard, recognizing that what is considered new will vary from organization to organization. This overview will heighten your attention to these areas when we explore the specifics in Chapters 2–9. Our list of content that may be considered new includes:

- Annex SL structure

- Understanding the organization and its context

- Understanding the needs and expectations of interested parties

- Actions to address risks and opportunities

- Organizational knowledge

Details for each of these will be provided in the body of this book. Here, we provide a few high-level comments:

Annex SL structure—In ISO 9001:2008 the requirements are contained in clauses 4–8. In ISO 9001:2015 the requirements are contained in clauses 4–10. The new structure of the requirements clauses is made mandatory by the ISO Directives. It is intended for use by all ISO management system standards (MSSs) with the intent expressed by ISO headquarters (which is arguable) that a common structure for MSSs will facilitate compliance by organizations.

Some will view the new structure as "new." Others will shrug their shoulders and comment that this new structure is nothing more than "reshuffling the deck." After you work with 2015 you can decide whether the new structure makes sense and adds any value. Your authors have their own opinion, which is discussed in Chapter 1.

Understanding the organization and its context—These words are new to ISO 9001 and are an attempt to require the organization to consider its strategic direction so it can determine issues of real or potential impact and plan and deploy processes and controls to manage such issues. Although the words may be new, many organizations have strategic and tactical planning processes in place to address such a requirement. If such processes do not exist, 2015 provides an incentive to consider implementation. To provide a frame of reference for this concept, Figure 1.1 provides an overview picture of the processes of the QMS and the extent of what may be included in the appraisal of the context of the organization.

Understanding the needs and expectations of interested parties—Interested parties? What does this mean? In prior editions of ISO 9001 there was strong concern for meeting customer, statutory, and regulatory requirements but no reference to any other parties. However, given the requirement related to the organization and its context, there is now a need to consider other relevant interested parties in addition to the customer. Much more on this later in Chapters 3 and 5.

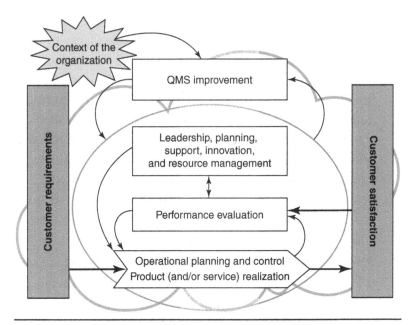

Figure 1.1   Model of a process-based QMS.

Actions to address risks and opportunities—Although we view the requirements around the concept of risk as vague, the intention of introducing risk into ISO 9001 is to encourage making preventive action an explicit element of the QMS. For many organizations, this will not be regarded as new. For others, new processes and thinking may be required. Although the wording in the standard may be vague and non-prescriptive to some, the concept is well worth attention if it has been ignored in the past. Chapter 5 discusses this in detail.

Organizational knowledge—Knowledge? A "new" concept? This requirement, although vague, could have been viewed as an implicit requirement of clause 6 in ISO 9001:2008 by some organizations. For others, the concept and the processes required to meet explicit and implicit requirements in 2015 will be new initiatives.

There are no requirements in clauses 4–10 to consider processes related to innovation. It is mentioned in the introduction and in a note to one of the requirements in the improvements clause (clause 10). We bring your attention to the decision to remove requirements to consider innovation processes since we believe that innovation is and will continue to be an essential component of a QMS. Even though it is not an explicit requirement of ISO 9001:2015, many organizations will consider including innovation requirements in their QMS from a sustainability viewpoint.

In addition to the concepts that may appear to be new in ISO 9001:2015, there has been a change in some of the "old" and "familiar" terminology that may be problematic for some organizations. Primary examples include:

- "Documented information" versus "documents and records"

- "Products and services" versus "products"

- Generic language

- "New" words or old words with tweaked definitions

An overview of these changes in terminology and a brief commentary on the rationale are as follows:

"Documented information" versus "documents and records"—In 2015 the words "document(s) and record(s)" do not appear, having been replaced by "documented information." This change in terminology was dictated by the imposition of Annex SL on ISO 9001 with the thinking that "documented information" was a preferred way of referencing both documents and records. This terminology may be problematic in some organizations, especially those operating in regulated marketplaces. These words will be discussed in Chapters 2–9.

"Products and services" versus "products"—In 2015 the word "products," which has been in ISO 9001 since

1987, has been replaced with the term "products and services," under the assumption that "products" was not generic enough and not friendly to operations other than manufacturing. Many disagree with this reasoning (including your authors), but we are stuck with the new term.

Generic language—Wherever possible throughout the 2015 standard there has been an effort to use language that is as generic as possible, under the assumption that, in the past, ISO 9001 has been biased toward manufacturing organizations.

"New" words or old words with tweaked definitions—Users of ISO 9001:2015 will need to obtain a copy of ISO 9000:2015 to ensure a precise understanding of the words and how auditors will understand such words. Examples of some of the words that may be new or have tweaked definitions whose formal ISO definitions should be understood include:

— Organization

— Interested party

— Risk

— Documented information

— Outsource

— Continual improvement

— Corrective action

— Correction

— Context of the organization

— Objective evidence

— Concession

ISO 9000:2015 is a normative reference in ISO 9001:2015, meaning definitions in ISO 9000 shall be considered as requirements.

Other aspects of ISO 9001:2015 that may appear to be new include:

- Less prescriptive requirements

- Integrating the process approach and systems thinking

- Leadership emphasis

These aspects of 2015 are examples of either departures from the previous editions of ISO 9001 or areas of increased emphasis. Examples of how these aspects of the 2015 standard are integrated into the requirements are as follows:

Less prescriptive requirements—An attempt was made in the 2015 edition to minimize prescriptive requirements. Whereas the 2008 edition contained requirements for a quality manual and a management representative, the 2015 edition has eliminated these. Also, there are many places where documented information (i.e., procedures) is not explicitly required. Remember that ISO 9001:2015 requires the organization to determine what documented information should be maintained and/or retained. So, the lack of a specific requirement in a particular clause does not mean the organization need not give careful consideration to determining what documented information is needed to ensure and demonstrate conformity. Users are also cautioned to consider auditing implications when deciding whether documented information (i.e., a procedure or a record) should be required.

Integrating the process approach and systems thinking—Although the process approach has been an explicit concept embedded in ISO 9001 since 2000, in 2015 the standard clarifies and emphasizes the requirement to apply the process approach to the QMS.

Leadership emphasis—Leadership receives much more explicit emphasis in 2015 than in past editions. Clause 5.1

states 10 specific items as requirements for management to demonstrate leadership and commitment to the QMS.

HOW TO GET STARTED

The latest edition of ISO 9001 contains enough new content, either in the form of new requirements or new terminology or in old concepts that have been repackaged, to prompt even organizations with a sophisticated QMS to pause before considering any changes. The latest edition provides an opportunity to review where your QMS is and where it needs to be to meet the strategic objectives of your organization over the next several years. The objective is to exist tomorrow and to prosper during the five or more years that this edition of ISO 9001 will be relevant. Figure 1.2 illustrates a relatively simple process for accomplishing this.

Although it is not a requirement of ISO 9001, we recommend the foundation of your QMS as a starting point. Ensure that it is firm. Such consideration should include a review of at least the following:

- Current relevance of the vision and mission statements of your organization

- The quality management principles (see ISO 9000:2015)

- The objectives of your organization

- The strategic plan of your organization

Without a firm foundation the QMS cannot be a vital element of the management of the organization. At best it will be viewed as a necessary evil; at worst, it will be an impediment to what is considered the main purposes of the organization. Your challenge is to ensure that the QMS is viewed as an equal partner in contributing to the success of the organization—just like finance, marketing, sales, production, engineering, and

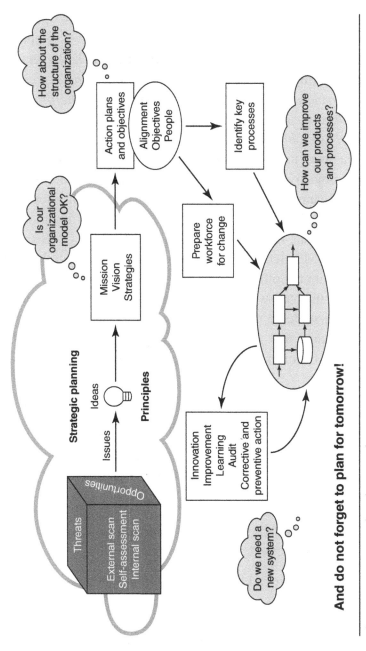

Figure 1.2 Internal and external context.

all the other vital parts of the organization. The pursuit of the well-defined and well-articulated quality objectives is not and should not be viewed as extra work. **It is the work of everyone in the organization.** *The quality professionals are responsible for ensuring that the QMS is structured so that the processes are indeed viewed as a contributor to organizational success and not as an impediment.*

The work to review and update, as appropriate, the mission, vision, objectives, and strategic plan of an organization is not easy and cannot be done by one individual. It requires top management involvement and, in most cases, several iterations. It is essential that this work precede a review of the QMS so that required QMS changes will be consistent with, integral to, and aligned with the other processes of the organization.

Once the hard work of updating the mission, vision, objectives, and strategic plan of the organization is addressed, then understanding and addressing the requirements of ISO 9001: 2015 can be initiated in earnest. We suggest the following sequence:

- Understand the meaning *and* intent of every clause in the standard

- Determine how to address the content of the standard that may be new, including the clauses related to:

  — Understanding the organization and its context

  — Understanding the needs and expectations of interested parties

  — Actions to address risks and opportunities

  — Organizational knowledge

- Develop a strong level of managerial intensity around the project

- Perform a gap analysis on the rest of the requirements of ISO 9001:2015 to determine what processes, if any, require modification to meet the explicit requirements of your organization and the standard

- Ensure consistency with the mission, vision, objectives, and strategic plan of the organization

- Make modifications, as appropriate

- Update documented information (to use the 2015 language for documentation)

- Provide training, as appropriate, for all processes that have been modified

- Audit all modified processes

- Iterate forever

This approach may take weeks, months, or longer to plan and implement. The result will be a QMS that not only complies with the requirements and intent of ISO 9001:2015, but also is an integral element of ensuring the sustainability of the organization.

There are easier approaches to attack the processes of an organization in order to pass a certification audit, but the result has a high probability of being a QMS that is reluctantly tolerated, does not add value, and is likely to exhibit inconsistent implementation.

Chapters 2–12 of this book will provide you with much of the information you require to perform a gap analysis to make appropriate modifications and to deploy compliant and value-adding processes.

The journey should be interesting as well as challenging. Enjoy your interaction with ISO 9001:2015.

2

Clauses 0–3—Introduction, Scope, Normative References, Terms and Definitions

Men's fundamental attitudes towards the world are fixed by the scope and qualities of the activities in which they partake.

—JOHN DEWEY

INTRODUCTION

Clause 0.1—General

Clause 0.1 of ISO 9001:2015 is similar to clause 0.1 of ISO 9001:2008. The material in the introduction is called "informative" in ISO language, meaning that it does not form part of the requirements of ISO 9001:2015. It exists to provide context, general understanding, and background.

This clause discusses the intent of ISO 9001:2015 and the potential benefits to an organization of implementing a QMS based on this standard. A summary of the potential benefits indicated is as follows:

- Provide products and services that are consistent in meeting requirements

- Enhance the ability to achieve customer satisfaction

- Address risks and opportunities associated with the organization's context and objectives

- Demonstrate conformity to specified QMS requirements

It also indicates that ISO 9001 is not intended to require or imply a requirement for uniformity of structure of a QMS, a need to align the documented information of an organization with the clause structure of this International Standard, or the need to use the same terminology used in the International Standard.

The introduction states that ISO 9001:2015 is based on the process approach, which incorporates the Plan-Do-Check-Act (PDCA) cycle and risk-based thinking. Both of these concepts are described in the requirements clauses of the International Standard and discussed in several places in Chapters 3–9 of this book.

To ensure clarity of understanding, the introduction "defines" terminology that is used throughout the International Standard. Following are the verbal forms used and what they mean:

- "Shall" indicates a requirement

- "Should" indicates a recommendation

- "May" indicates a permission

- "Can" indicates a possibility or a capability

Clause 0.2—Quality Management Principles

ISO 9001:2015, like the previous editions of ISO 9001, was developed with the quality management principles (QMPs) that are given in ISO 9000:2015 as a basis. While the QMPs help form the foundation of ISO 9001, they do not appear in ISO 9001:2015 and are not part of the requirements. The principles as they appear in ISO 9000:2015 are as follows:

- Customer focus

- Leadership

- Engagement of people

- Process approach

- Improvement

- Evidence-based decision making

- Relationship management

In addition to stating the principles, ISO 9000:2015 also includes a rationale of why the principle is important, a few examples of benefits associated with the principle, and examples of typical actions that an organization can take to improve performance when applying the principle.

Since the QMPs are a foundational element of the QMS of an organization, it is our strong recommendation that they be reviewed along with the vision and mission statements of the organization when contemplating processes to meet the requirements and intent of ISO 9001.

Clause 0.3—Process Approach

The process approach has been one of the QMPs for many years and a formal and explicit component of ISO 9001 since it was incorporated into the 2000 edition of ISO 9001. In 2015 the process approach receives emphasis by including explicit requirements (see clause 4.4) to be addressed when developing, implementing, and improving the effectiveness of a QMS. Clause 0.3, in the nonnormative section of the standard (i.e., this section does not contain requirements), emphasizes the importance of the process approach integrated with risk-based thinking (see clauses 0.3.3 and 6) and with the holistic concepts of systems thinking.

Figure 1 in the introduction provides a model of a single process. Our preferred simple model for a process is shown in

Figure 2.1, which shows the classic representation of a process as a set of interrelated or interacting activities that transforms inputs into outputs (the formal ISO 9000:2015 definition of a process). Multiple interacting processes and resource inputs almost always exist but for simplicity are not shown in Figure 2.1.

Figure 2 in the introduction attempts to provide a model of process interactions in the context of the PDCA model. Our preferred model of all the elements and interactions of ISO 9001:2015 is shown in Figure 1.1 in Chapter 1. This figure also incorporates the concept of the "context of the organization" that is introduced into the International Standard in clause 4.1, along with a visual portrayal of the processes of the QMS.

Clause 0.3 emphasizes the importance of the process approach and consideration of every process of the QMS for improvement based on measurement or monitoring of performance and analysis of data. Clause 0.3 also introduces the concepts of risks and opportunities. Inclusion of these activities in the introduction is intended to warm up users of the standard to the current approach incorporated in the requirements for these activities. The primary "new" point being made in the introduction is that risk should be considered from a positive

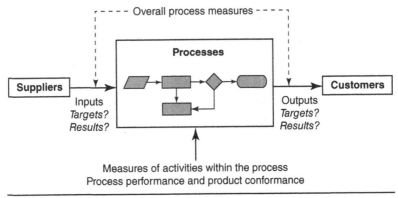

Figure 2.1   Process: A group of interrelated activities and related resources that transforms inputs into outputs.

viewpoint (i.e., "How can we approach attracting new customers?" and "What is the risk of not attracting new customers?"), not just from the normal negative viewpoint (i.e., "What can go wrong?").

The introduction also includes material on the PDCA cycle, as mentioned earlier, but use of the PDCA cycle is not a requirement in ISO 9001:2015. It is one of many approaches to process management, control, and improvement that can be applied to individual processes of the QMS. Quality practitioners have been employing PDCA for many years when and where it is appropriate, even long before the initial publication of ISO 9001 in 1987. We leave it to the curious reader to ascertain why ISO 9001:2015 devotes about a page to describing the PDCA cycle and attempts to integrate it with the entire QMS, which we believe is misleading and also can be counterproductive. We have not found this simplistic model to be adequate for managing an entire QMS.

Clause 0.4—Relationship with Other Management System Standards

Clause 0.4 applies the framework developed by ISO to improve alignment among its International Standards for management systems (i.e., Annex SL). The ISO assertion (we believe without merit) for imposing the common format on all MSSs was that the common structure would facilitate an organization in aligning or integrating its QMS with other management systems (e.g., environmental). We believe, having implemented and achieved multiple management system certifications in an organization, that the attempt to force a common structure is misguided at best and will create much more non-value-added effort than it will save. But we must live with the decisions that have been dictated by ISO management.

Clause 0.4 also references:

- ISO 9000:2015 *Quality management systems—Fundamentals and vocabulary*, which provides essential background for the proper understanding and implementation of ISO 9001:2015

- ISO 9004:2009 *Managing for the sustained success of an organization—A quality management approach,* which provides guidance for organizations that choose to progress beyond the requirements of this International Standard

ISO 9000 is a normative requirement for ISO 9001, and all users of ISO 9001 need to obtain and use ISO 9000. It is indispensable if certification is the organization's objective—and very helpful even if it is not.

ISO 9004:2009 is a useful document for an organization that desires to expand the breadth and depth of its QMS. ISO 9004:2009 is just starting its "update" process, so it may be prudent to obtain a copy of this standard after the ISO 9004:2009 version is updated, which may occur in 2017–2018.

Several sector-specific QMS standards have been developed. It is anticipated that these documents will be updated to use ISO 9001:2015 as their core requirements. These sector-specific MSSs specify additional requirements pertinent to the sector-specific marketplaces. Sector-specific QMSs are addressed in Chapter 13.

To provide users of ISO 9001:2015 with more reference information, the introduction mentions that Annex B provides details of other International Standards on quality management and QMSs that have been developed by the ISO Quality Management Technical Committee (TC 176).

The final nugget of information in the introduction is the statement that a matrix showing the correlation between the clauses of ISO 9001:2015 and ISO 9001:2008 can be found on the ISO/TC 176/SC 2 open-access website at http://www. iso.org/tc176/sc02/public. Support documentation related to understanding and deploying processes to comply with its requirements is available at the ISO/TC 176/SC2 public website at http://isotc.iso.org/livelink/livelink/open/tc176SC2public.

SCOPE

The scope clause in ISO 9001:2015 is almost identical to the scope included in ISO 9001:2008. It is "informative" in ISO language, meaning that it does not form part of the requirements of ISO 9001:2015. It exists to provide context, general understanding, and background and to indicate why organizations should use it. This can be summarized as follows:

- To demonstrate the organization's ability to consistently provide product or service that meets customer and applicable statutory and regulatory requirements, and the organization's own internal requirements

- To enhance customer satisfaction through the effective application of the system, including processes for improvement of the system and the assurance of conformity to customer and applicable statutory and regulatory requirements

The scope also states that ISO 9001:2015 is intended to be generic and applicable to all organizations, regardless of type, size, and product or service provided.

In a note in the scope, users are reminded that the terms "product" and "service" in the requirements only apply to products and services intended for, or required by, a customer. This note will cause some head-scratching by users of the International Standard when they attempt to understand the requirements in clause 4.2 related to "the requirements of relevant interested parties" and processes to conform with the requirements as required in clause 6. These issues will be covered in Chapters 3 and 5 of this book.

ISO 9001:2008 included text in the scope requiring justification for requirements that were excluded from the QMS of the organization. ISO 9001:2015 requires conformance to all applicable requirements (see Chapter 3).

NORMATIVE REFERENCES

ISO 9000:2015, *Quality management systems—Fundamentals and vocabulary*, is indicated to be a normative reference, which means that the terms and definitions in ISO 9000:2015 shall be applied and used when developing the QMS, creating documented information, and deploying processes related to ISO 9001:2015. As we previously mentioned, users of ISO 9001:2015 need to obtain and use ISO 9000:2015.

TERMS AND DEFINITIONS

ISO 9000:2015, *Quality management systems—Fundamentals and vocabulary*, is the source for all defined terms used in ISO 9001:2015.

3

Clause 4—Context of the Organization

Nothing true or beautiful or good makes complete sense in any immediate context of history; therefore we must be saved by faith. Nothing we do, however virtuous, can be accomplished alone.

—REINHOLD NIEBUHR, 1952

INTRODUCTION

Clause 4 incorporates concepts and words that will be new to many who are familiar with the prior editions of ISO 9001.

Some of the new concepts introduced in clause 4 include:

• Understanding the organization and its context

• External and internal issues

• Interested parties that are relevant

Four subclauses within clause 4—4.1, 4.2, 4.3, and 4.4—elaborate on what these concepts and words require.

Clause 4.1 contains requirements for an organization to determine external and internal issues that can impact on and are relevant to the purpose and strategic direction of the organization. The organization shall also monitor and review information related to relevant external and internal issues.

Clause 4.2 introduces the concept of an organization determining what interested parties are relevant to the organization and, therefore, to its QMS. It further indicates the requirement that the organization determine the requirements of relevant interested parties and directs the organization to monitor and review information about relevant interested parties.

Clause 4.3 digresses a little to define a requirement that the organization address the boundaries and applicability of its QMS so it can provide a full description of the QMS scope. This clause also indicates areas that the organization shall consider when it determines its scope, such as the external and internal issues that are determined (as required in clause 4.1) and the requirement of the relevant interested parties that were identified in clause 4.2.

The clause requires consideration of the organization's products and services, which, we believe, should be the primary consideration of the organization. This clause also addresses what, in the past, was the issue of exclusions, and it is clear that if any requirement of ISO 9001 can be applied to the products and services within the scope of the organization, it shall be applied. The requirement cannot be excluded. It further states that if any requirement of ISO 9001 cannot be applied, the "exclusion of compliance" shall not affect either the ability or the responsibility of the organization to ensure the conformity to requirements of its products and services.

The final requirement incorporated in clause 4.3 is that the scope shall be available and maintained and shall state the products and services to which the QMS applies as well as a justification for any exceptions the organization makes for conforming to any of the ISO 9001 requirements.

Clause 4.4 contains requirements for the organization to establish, implement, maintain, and continually improve a QMS, including the processes needed and their interactions. The requirements stated are consistent with what was contained in previous editions of the standard, but in 2015 the requirements go a little further in emphasizing attention to the processes needed and the interaction of processes. The clause also states eight areas related to the needed processes that the organization shall determine and address.

Clause 4.4 also has a requirement related to documented information that shall be maintained to support process operation and the retention of documented information to provide confidence that the processes are being implemented as planned.

This introduction to clause 4 is intended to provide a framework for the more detailed discussion of each of the subclauses that follows.

In summary, the four subclauses require the organization to understand external and internal issues relevant to its:

* Purpose

* Strategic direction

* Interested parties

* Processes

* Ability to achieve the intended results of its QMS

The organization is also required to monitor and review information on internal and external issues.

CLAUSE 4.1—UNDERSTANDING THE ORGANIZATION AND ITS CONTEXT

What Does the Clause Say?

Clause 4.1 introduces the concept of requiring an organization to think at both the strategic and tactical levels when it develops and deploys its QMS. Neither big-picture thinking nor detailed analysis is sufficient by itself. The clause mandates consideration of internal and external issues affecting the ability of the organization to achieve its intended results. The requirements also state the organization shall monitor and review the issues it considers to be relevant to its purpose. Notes to this clause are provided for clarity but do not contain requirements. (This is true for all notes to a requirements standard.) In this case the notes suggest the organization consider issue areas such as the marketplace, competition, the environment, society, and the economy as well as issues that relate to the culture of the organization and its performance.

What Does the Clause Mean?

The requirements for understanding the organization and its context mean the organization needs to know itself and the external organizations and factors affecting it or that it can affect. Achieving such understanding can result from activities such as performing competitive analysis, assessing the organization's reliance on existing and emerging technology, and evaluating its impact on the relevant interested parties. Such activities and assessments are elements of the strategic and tactical planning for the organization overall and for its associated QMS. They also form a context for developing, implementing, maintaining, changing, and improving the QMS.

This subject is a normal topic for top managers and is interrelated with clause 5.1.1, *Leadership and commitment—General*, which requires top management to ensure the quality policy and quality objectives are compatible with the organization's strategic direction and context. This is a key top management role in the development of the QMS.

Consider the Potential Interactions as They Apply to Your QMS

Clause 4.1 may have direct or indirect interactions with several other clauses. In particular, the organization should consider the contents of clauses 4.2, 5.1, 6.1, 6.2, 7.3, 8, 9, and 10 when considering processes to conform with the requirements of clause 4.1.

Implementation Tips

What is meant by "determine external and internal issues"? What should be considered? How far should the organization go?

One approach we suggest is to have a process or processes for deciding what to consider and why. This makes good business sense for several reasons, including (1) to formalize the process to ensure it is followed, (2) to preclude going overboard on the determination of pertinent external and internal issues, and (3) to preempt disputes with external auditors regarding compliance (if certification is an organizational objective). The notes to clause 4.1 provide explanatory guidance (as indicated earlier).

For example, the organization could develop a list of areas where issues could exist, and could perform periodic evaluations of any factors that exist or are emerging that may impact meeting requirements.

Examples of methods commonly used to determine internal issues include:

- Internal audit results and self-assessment results

- Analysis of quality cost data

- Analysis of technology trend information

- Competitive analysis

- Results of customer reviews, audits, complaints, and feedback

- Actual versus intended internal values and culture

- Organizational performance

- Best practices of the organization and comparisons with industry benchmarks

- Employee satisfaction data analysis

Self-assessment is an underutilized but very powerful process for identifying internal issues. It should be given serious attention.

Self-assessments can be complex, using criteria such as those of the Malcolm Baldrige National Quality Award, the European Foundation for Quality Management (EFQM), or the American Society for Quality (ASQ) guidelines for performing a self-assessment of a QMS. Assessment can also be simplified by using the seven QMPs as a guide. It is up to the organization to determine how detailed the analysis should be and the follow-up action, monitoring, and review that are needed.

External issues can be found through a variety of techniques such as analysis of:

- The economic environment and trends

- International trade conditions

- Competitive products and services

- Opportunities and conditions related to outsourcing

- Technology trends

- Raw material availability and prices

- Benchmarking best-in-class performers both within and outside the current marketplace

This new requirement should provide the organization with an opportunity to expand the breadth and depth of its QMS and its integration into both the strategic and tactical management of the organization. It should also facilitate alignment of objectives throughout the organization.

Questions to Ask to Assess Conformity

- Is there evidence the organization has determined relevant external and internal issues?

- Is there evidence the organization has monitored and reviewed information for relevant external and internal issues?

Definitions (Refer to ISO 9000:2015)

- Context of the organization
- Organization
- Quality management system

Considerations for Documented Information to Be Maintained and/or Retained

- Processes to determine external and internal issues relevant to the purposes of the organization
- Processes to monitor and review information about external and internal issues

CLAUSE 4.2—UNDERSTANDING THE NEEDS AND EXPECTATIONS OF INTERESTED PARTIES

What Does the Clause Say?

Clause 4.2 introduces a requirement to determine parties that have an interest in and are relevant to the products and services of the organization and to determine their requirements. It further requires the organization to monitor and review the information about these interested parties and their relevant requirements.

The concept of "interested parties" is new to ISO 9001:2015. One of the reasons stated for its inclusion is that there are parties beyond the direct customer that can impact the QMS of the organization to ensure it meets its objectives.

What Does the Clause Mean?

The actions related to interested parties will broaden the viewpoint of an organization. It may not be sufficient for an organization to just meet the requirements of customers. Depending on the nature of the products and services of the organization, it may be necessary to consider the requirements of interested parties that can have an impact on the organization, such as society, regulatory bodies, suppliers, and "special interest groups" (e.g., environmentalists). This broadening of the sphere of concern may be either confusing or mystifying (or both) to organizations for several reasons. Recall the concept found in the introduction to the International Standard, which states an organization can use the International Standard to assess the ability to consistently meet customer, statutory, and regulatory requirements as well as its own requirements. Similar words are in the scope as well. There is no front-end mention of "interested parties." However, clause 4.1, as mentioned earlier, requires the organization to "determine external and internal issues," clause 4.2 requires the organization to determine relevant interested parties, and clause 6.1 requires the organization to consider the issues that are determined.

This means there may be substantial variation from organization to organization and from auditor to auditor. Organizations may have difficulty discerning how to approach conformity with these requirements.

This clause means the organization shall have a process to decide what interested parties, if any, are relevant; to determine how it shall monitor and review the information about these interested parties and their relevant requirements; and to retain documented information of process implementation.

Words like "relevant," "determine," "monitor," and "review" in this requirements standard have the potential to introduce ambiguity and issues with external auditors.

Consider the Potential Interactions as They Apply to Your QMS

Clause 4.2, like clause 4.1, may interact with several other clauses, either by direct process flow or by much more subtle connections. In particular, the organization should consider the contents of clauses 4.1, 5.1, 6.1, 6.2, 9, and 10 when considering processes to comply with the requirements of clause 4.2.

Implementation Tips

We recommend organizations define and deploy processes to determine whether interested parties are impacted by the activities of the organization, and if there are such parties, to require consideration of actions to ameliorate or eliminate the extent of the impact.

We also recommend that while considering the real or potential impact of interested parties, organizations maintain the primary focus on meeting the requirements of their customers but not ignore the requirements of relevant interested parties. Figure 3.1 shows one example of potential interested parties that may be relevant to an organization.

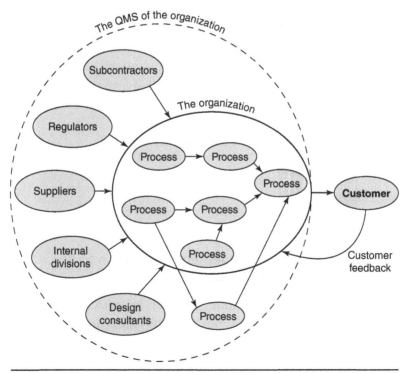

Figure 3.1 Example of potential interested parties.

For example, suppliers need to deliver conforming components or subassemblies on time. Society may be concerned with impacts on the community. Shareholders are concerned with the valuation of the stock price for a company that may be affected by operating efficiency, effectiveness, and the achievement of market share expansion.

It is the organization's sole prerogative to determine the relevance of interested parties and their requirements, and what and how they will monitor and review the information about interested parties. It may present a challenge to the leaders of the organization to ensure alignment within the organization regarding how to address the consideration of interested parties.

Questions to Ask to Assess Conformity

- What is the process for determining whether interested parties are impacted by the activities of the organization?

- What is the process for determining the requirements of interested parties?

- How is the monitoring and reviewing of the information about interested parties and their relevant requirements conducted?

Definitions (Refer to ISO 9000:2015)

- Interested party

- Monitor

- Requirements

Considerations for Documented Information to Be Maintained and/or Retained

- Processes to determine whether interested parties are impacted by the activities of the organization

- Processes for determining the requirements of interested parties

CLAUSE 4.3—DETERMINING THE SCOPE OF THE QUALITY MANAGEMENT SYSTEM

What Does the Clause Say?

Highlights of the requirements for determining the scope of the organization's QMS are as follows:

- Determine the boundaries and applicability of the QMS

- When determining the QMS scope, consider:

  - The external and internal issues determined during analysis of the context of the organization

  - The requirements determined during analysis of the organization's relevant interested parties (see 4.2)

  - The organization's products and services

- Application of a requirement is always required where it is within the scope of the QMS (i.e., any exceptions need rigorous control)

- Any requirement the organization cannot apply shall not affect its ability or responsibility to ensure conformity of product or service

- The scope is to be available

- The scope shall be maintained as documented information

- If any requirement is deemed not applicable to the scope of the QMS, justification for this decision shall be included in the QMS scope

What Does the Clause Mean?

The requirements for determining the scope of the QMS emphasize the importance of determining the boundaries and the application of the QMS. Areas to be considered when determining the scope include:

- External and internal issues (see 4.1)

- The requirements of relevant interested parties (see 4.2)

- The organization's products and services

The documented information of the scope shall state:

- How it is maintained

- Its availability

- The products and services embraced

- Justification for any situation(s) where any requirement cannot be applied

Consider the Potential Interactions as They Apply to Your QMS

Clause 4.3 may interact with clauses 4.1, 4.2, 6, 9, and 10.

Implementation Tips

We recommend treating QMS scope development as a process. The QMS scope may be very static, or it may need to be subject to frequent change in a very dynamic organization. In the latter case, the organization should consider maintaining documented information on how this process is to be conducted.

Scope statements should be as brief and succinct as feasible and still conform to the requirements. Note that the scope is to be maintained as documented information and be available, but there is no requirement for having documented information of the process to develop the scope or for when and how the scope is updated, although we believe it is prudent to have documented information describing the development and maintenance of the scope of the organization.

Questions to Ask to Assess Conformity

- How are the boundary conditions of the QMS established?

- How are external and internal issues considered?

- How are the requirements of relevant interested parties considered?

- What is the process and the output of the process to consider the products and services to be included in the scope of the organization?

- Are nonapplicable requirements justified in the scope?

Definitions (Refer to ISO 9000:2015)

- Conformity

- Interested party

Considerations for Documented Information to Be Maintained and/or Retained

- Scope development and maintenance process

- How the scope is made available

- Justification of all non-applicable requirements

CLAUSE 4.4—QUALITY MANAGEMENT SYSTEM AND ITS PROCESSES

What Does the Clause Say?

The organization shall determine the processes needed for the QMS and the application of the processes throughout the organization. In doing this it shall determine:

- The inputs required and outputs expected from processes

- The sequence and interaction of processes

- The criteria, methods, measurements, and performance indicators to ensure effective operation and control of the processes

- The resources needed

- The availability of resources

- The responsibilities and authorities for these processes (e.g., a process owner)

- The risks and opportunities (see clause 6)

- The implementation of actions to address risks and opportunities

- Methods for monitoring, measuring (as appropriate), and evaluating processes and, if needed, the control of changes to processes to ensure outputs meet requirements

- Processes to explore opportunities and responsibilities for improvement of the processes and the QMS

The organization shall also maintain documented information as appropriate to support process operations, and retain documented information to the extent necessary to have confidence that processes are being carried out as planned and under controlled conditions.

What Does the Clause Mean?

Many of the requirements are similar to the previous version of ISO 9001. The new "wrinkles" include either an explicit or implicit requirement to address interested parties, the context of the organization, and internal and external issues when planning, controlling, and improving the processes of the QMS. In most other respects, the requirements in the 2015 edition are similar to the 2008 edition.

Consider the Potential Interactions as They Apply to Your QMS

Clause 4.4 may interact with clauses 5.1, 6, 7, 8, 9, and 10.

Implementation Tips

Before contemplating the QMS and its processes, the organization should consider the foundations of the QMS (see Figure 3.2), including a review of and updating, as appropriate, the mission and vision of the organization, the quality policy, the quality objectives, the alignment of the policy and objectives with the overall vision and mission of the organization, and established strategic and tactical plans.

It is also important to explore and understand basic customer needs and requirements. This understanding is key input to the development of the processes of the QMS. Most organizations will already have developed and deployed at least some, if not most, of the processes that are needed for an effective QMS that will meet customer needs and requirements.

A next step could be to identify these existing processes and determine what other processes, if any, must be developed

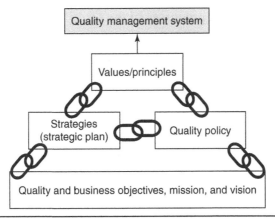

Figure 3.2   The foundation underpins the system.

for the QMS to be effective and conform to ISO 9001:2015. For several reasons it is helpful to map at least the key processes. In addition to the individual processes that are required for an effective QMS, the organization must determine how all of the processes relate to each other (i.e., the interfaces and interactions of processes). Understanding these relationships is important because problems often occur where processes interface or interact. Process development should include addressing all the areas indicated earlier as requirements, with particular attention to process inputs, outputs, performance indicators, risk assessment, interactions, improvement, and controls.

See "How to Get Started," in Chapter 1, for a summary of the activities organizations can consider when implementing a QMS in conformance with ISO 9001:2015 requirements.

Questions to Ask to Assess Conformity

- Have the processes needed for quality management been identified?

- Have sequence and interaction of these processes been determined?

- Have criteria and control methods been determined for control of the processes in the QMS?

- Is documented information available to support the operation and monitoring or measuring of the processes?

- Are processes measured, monitored, and analyzed, with appropriate actions taken to achieve planned results and improvement?

- Is the QMS established, documented, implemented, maintained, and improved?

- Has provision been made to ensure control of QMS processes that are performed outside the organization?

Definitions (Refer to ISO 9000:2015)

- Management system

- Organization

- Process

- Quality

- Quality management system

- Requirement

Considerations for Documented Information to Be Maintained and/or Retained

- Sufficient to support the operation and control of processes

- Sufficient to provide confidence that the processes are being carried out as planned

4

Clause 5—Leadership

Leadership and learning are indispensable to each other.

—JOHN FITZGERALD KENNEDY, 1963

INTRODUCTION

The requirements for leadership will be viewed by many as a significant enhancement from earlier editions of ISO 9001. Others will mumble that it is about time there are more rigorous requirements to intensify management engagement. Still others will observe the latest edition does not go far enough. Thus there is a spectrum of opinions regarding the 2015 leadership content, and the general consensus is that improvement of the content has been achieved.

In preparation for a more detailed discussion of this clause, let us consider the obvious "enhancements" of requirements, which are:

- Expansion of the specific requirements (see 5.1.1)

- Ensuring that the policy and objectives are compatible with the strategic direction

- Expansion of the quality policy requirements (communicated, understood, and applied)

- Integration of the QMS into business processes

The leadership clause has three subclauses: 5.1, *Leadership and commitment* (with two subclauses); 5.2, *Policy* (with two subclauses); and 5.3, *Organizational roles, responsibilities, and authorities.*

WHAT DOES CLAUSE 5.1 SAY?

5.1.1 Leadership and Commitment—General

The organization's top managers are required to demonstrate their leadership and commitment to the QMS by carrying out specific activities that include at least the following:

- Taking accountability for QMS effectiveness

- Ensuring the following:

  — Establishment and maintenance of the quality policy

  — The policy is compatible with the organization's context and strategic direction

  — The policy is communicated, understood, and applied

  — Integration of the QMS into business processes

  — Availability of necessary resources for the QMS

  — The QMS achieves intended results

  — Customer requirements are determined and met

  — Applicable statutory and regulatory requirements are determined and met

  — Risks and opportunities that may affect conformity of products and services are determined and addressed

— Risks and opportunities related to the ability to enhance customer satisfaction are determined and addressed

— There is a focus on consistent delivery of products and services that meet customer requirements

— Engaging, directing, and supporting contribution to QMS effectiveness

— Promotion of continual improvement

— Understanding of the process approach in the organization

— Communication of the importance of effective quality management and conformity to the QMS requirements

— Supporting other managers to demonstrate leadership in their area of responsibility

As was mentioned in the introduction to this clause, the explicit requirements related to top management have been expanded. It should be noted that top management has another role related to its involvement with the QMS. This role is participation in management review, as addressed in Chapter 8 (which covers the requirements of clause 9).

5.1.2 Customer Focus

Top management also has a requirement to demonstrate leadership and a commitment to customer focus by ensuring:

• Customer and any applicable regulatory requirements are determined and met

• Risks and opportunities that can impact conformity to requirements or customer satisfaction are determined and appropriate actions are taken

• A continual focus is maintained on customer satisfaction

WHAT DOES CLAUSE 5.2 SAY?

5.2 Policy (Includes Clauses 5.2.1 and 5.2.2)

Top managers shall establish the quality policy, review it, and maintain it and ensure that it:

- Is appropriate to the organization's purpose and context

- Provides a framework for setting and reviewing quality objectives

- Includes commitment to meet requirements

- Includes commitment to improvement

- Is available as documented information

- Is communicated and understood

- Is available to relevant interested parties

WHAT DOES CLAUSE 5.3 SAY?

5.3 Organizational Roles, Responsibilities, and Authorities

Another role for top managers is to ensure that responsibilities and authorities are assigned, communicated, and understood for the following activities:

- Ensuring that the QMS conforms to ISO 9001 requirements

- Monitoring processes to ensure they deliver intended outputs

- Promoting customer focus everywhere in the organization

- Maintenance of the QMS when changes are planned and implemented

- Reporting on the performance of the QMS

- Reporting on opportunities for improvement
- Reporting on the need for changes to the QMS

WHAT DOES CLAUSE 5 MEAN?

Clauses 5.1, 5.2, and 5.3 contain requirements for top management regarding its leadership and commitment to:

- The QMS
- Maintaining customer focus
- Establishing and maintaining the quality policy
- Ensuring that the responsibilities and authorities for relevant roles are assigned, communicated, and understood

The requirements for leadership and commitment in clause 5.1.1 are explicit and clear. They are also extensive. Top managers need to be in control of the QMS through direct actions, where appropriate, or through delegation. The overall leadership of the QMS is not delegable, but in organizations of any real size, many components must be addressed by appropriate delegation. Care should be exercised to ensure that the assignments of responsibilities and authorities are clear and unambiguous. If two or more people share a responsibility or accountability, their individual roles need to be understood.

In the introduction we mentioned that there are a few requirements that may be considered as new, but we believe these requirements were implicit in the prior edition and they make sense. For example, it is reasonable to require that the quality policy not only be communicated and understood within the organization but also be applied.

It is also meaningful to review the QMPs that are one element of the foundation of ISO 9001. Leadership is presented in QMP 2, which is addressed in clause 2.3.2 of ISO 9000:2015. In 2015 the expectation is that the QMS shall require much more active participation by all levels of management. Emphasis is placed on full engagement of top management.

CONSIDER THE POTENTIAL INTERACTIONS AS THEY APPLY TO YOUR QMS

The leadership clause interacts with all the other clauses in ISO 9001:2015. Attention should be dedicated to interaction with the processes of clauses 4, 6, 7, 9, and 10 and in particular to activities related to:

- Clause 5.2 Quality policy

- Clause 6.2 Quality objectives

- Clause 8 Operation

- Clause 9 Performance evaluation

- Clause 10 Improvement

IMPLEMENTATION TIPS

If we take a minimalist view of clause 5, there are three major areas it addresses: (1) top management leadership and involvement, (2) the quality policy of the organization, and (3) roles and responsibilities.

A question that has plagued (and perplexed) quality professionals for a long time and has been a source of misplaced irritation is how to approach the first area. We say misplaced irritation because, rather than QA folks complaining about the lack of top management involvement, we should be looking inward at our own failure to establish effective communications with top management.

What should we be doing? Of course, all we can do here is provide a starting point for your own introspection to find processes that will work in your organization. One area to consider is how messages and requests for top management involvement are presented or communicated. Is interaction regarding issues related to quality management posited in the language of management? Do we talk about error rates or costs? Do we discuss acceptable quality levels (AQLs) or earnings per share (EPS)? Have we made quality costs visible in money terms? Do we understand how to integrate QA impact into a balanced scorecard? Are we knowledgeable about the impact of the London Inter-bank Offered Rate (LIBOR) or the Federal Reserve Board (FRB) on our organization? We need to embed in our minds that the language of top managers is money.

Quality professionals should consider development of processes that communicate requirements and needs in the language of management. By way of analogy, if an individual is in a bistro in a small village in southern France and is hungry, a successful attempt to obtain food would have a much higher probability of success if communication was conducted in either the French language or sign language than attempting to communicate in English.

Top management engagement in enhancing customer focus could be "packaged" in terms of enhancing market share and lowering various components of quality costs.

To address issues related to the formulation and deployment of quality policy, processes could be considered in terms of the quality policy being a vehicle for not only integrating internal activities with strategic and tactical plans but also enhancing marketplace perception of the organization and, in for-profit organizations, potential for favorable impact on marketing and sales.

Engagement and involvement of top management in ensuring clarity of roles and responsibilities could be facilitated by considering the financial and operational impact of processes. Impacts can be expressed in revenue terms, rather than in language used in defect reduction activities. Opportunities for improvement can be quantified in terms of expected return on investment. Such quantification in financial terms can motivate top managers to become and remain involved.

Top management should also be motivated to ensure clarity of roles, responsibilities, and allocation of resources. Providing personal leadership is the most important of the efforts of top management since the outputs from such efforts are corrective action, risk avoidance, cost reduction, process performance improvement, and enhanced customer satisfaction. All these outputs can have a positive impact on overall performance.

Documentation of processes for the communication and application of the quality policy can also be considered both to meet the overall requirement of ISO 9001:2015 and to improve internal performance.

QUESTIONS TO ASK TO ASSESS CONFORMITY

- Is there evidence that top management is demonstrating leadership and commitment with respect to the QMS as required by clause 5.1.1?

- Is there evidence that top management is demonstrating leadership and commitment to ensuring and maintaining customer focus?

- Has top management established a quality policy?

- Is the quality policy appropriate for the organization?

- Does the quality policy require the organization to satisfy applicable requirements?

- Has top management developed quality objectives that are consistent with the quality policy?

- Has the quality policy been communicated and is it understood and applied within the organization?

- Do top managers perform regular management reviews and assess opportunities for improvement?

- Is there evidence of top management commitment to improvement of the QMS?

- Does top management provide adequate resources, conduct regular resourcing reviews, and reallocate resources as needed?

- Is top management involved in the process to determine customer requirements and to ensure that they are met?

- Is there a process to ensure that employees understand the importance of meeting customer, regulatory, and statutory requirements?

- Is there evidence that top management ensures that the responsibilities and authorities for relevant roles are assigned, communicated, and understood within the organization in compliance with the requirements of clause 5.3?

DEFINITIONS (REFER TO ISO 9000:2015)

- Customer
- Customer satisfaction
- Improvement
- Management system
- Organization
- Output
- Quality
- Quality management system
- Quality objective
- Quality policy

- Requirement

- Risk

- Top management

CONSIDERATIONS FOR DOCUMENTED INFORMATION TO BE MAINTAINED AND/OR RETAINED

The only requirement for documented information in clause 5 is for the quality policy to be available as documented information.

Does this mean that an organization need not be concerned with any other documented information (i.e., procedures and records) to ensure the effective implementation of processes to ensure consistent conformity with clause 5 requirements? We believe the answer to such a question is no.

For addressing the numerous requirements in clauses 5.1.1, 5.1.2, 5.2.1, 5.2.2, and 5.3 it may be desirable (or necessary) to create and deploy several pieces of documented information (i.e., procedures) to articulate requirements for many processes, such as providing management review input and quality costs data gathering and analysis (which could be accomplished with several procedures), and for implementing balanced scorecard data gathering and reporting. Many other processes are also worth consideration for documentation even if there is no explicit requirement to ensure effective implementation (e.g., "How to integrate quality policy creation and deployment with strategic and tactical planning" and "How to ensure compatibility of the quality policy with objectives").

Although only one requirement for documented information is stated in clause 5, an organization will require many additional procedures or instructions to be created, maintained, and used to ensure consistent conformity with the requirements of this clause.

5

Clause 6—Planning

These unhappy times call for the building of plans that build from the bottom up and not from the top down.

—Franklin Delano Roosevelt, 1932

CLAUSE 6.1—ACTIONS TO ADDRESS RISKS AND OPPORTUNITIES (INCLUDING UNTITLED CLAUSES 6.1.1 AND 6.1.2)

What Does the Clause Say?

Clause 6.1 is interrelated with clauses 4.1 (external and internal issues) and 4.2 (interested parties). It requires consideration of issues relevant to the purposes of the organization (see 4.1) and determining the risks and opportunities that need to be addressed in planning the QMS to enable the organization to achieve intended results, prevent or reduce undesired effects, and achieve improvement. The planning activities are required to address how to integrate the actions into the QMS and how to evaluate effectiveness.

Clause 6.1 adds a statement that actions taken to address risks and opportunities shall be proportionate to the potential impact on conformity of products and services. A note to clause 6.1.2 offers suggestions for addressing risks and opportunities.

What Does the Clause Mean?

Clause 6.1 interacts with clauses 4.1 and 4.2. It contains new requirements that direct the organization to consider risks and opportunities as well as other interested parties when planning for the QMS. Figure 5.1 depicts how these areas can be integrated as elements of the overall QMS. Notice how both threats and interested parties (identified in internal and external scans) in the upper-left corner of the figure provide input into planning the QMS. Threats and opportunities should be assessed and addressed when planning the QMS, as they may impact both the organization and relevant interested parties. In addition, the organization is required to consider how to evaluate the effectiveness of the actions it plans to take to address risks and opportunities.

The last sentence of clause 6.1.2, "Actions taken to address risks and opportunities shall be proportionate to the potential impact on the conformity of products and services," is almost humorous to include in the standard. How can a standard mandate that an organization not take dumb actions?

CLAUSE 6.2—QUALITY OBJECTIVES AND PLANNING TO ACHIEVE THEM (INCLUDING UNTITLED CLAUSES 6.2.1 AND 6.2.2)

What Does the Clause Say?

Quality objectives shall be established at relevant functions and levels.

There is a laundry list of characteristics that shall be applicable to objectives:

- Consistent with the quality policy

- Measurable

- Based on applicable requirements

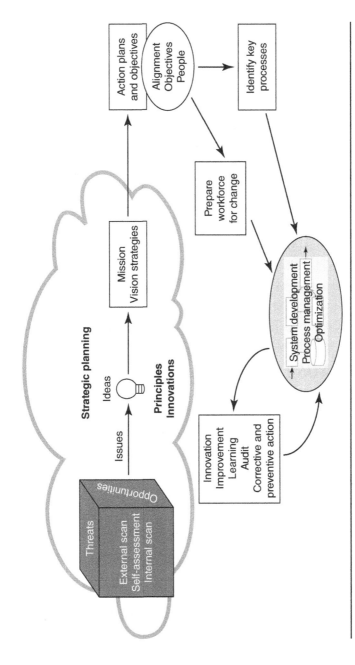

Figure 5.1 Keeping the QMS relevant.

- Relevant to conformity of products and services and to enhancement of customer satisfaction

- Monitored

- Communicated

- Updated, as appropriate

Retention of records (documented information) is required.

What Does the Clause Mean?

Clause 6.2 is self-explanatory. It means just what it says. The question that may be raised is, what is meant by "at relevant functions, levels and processes"? Our position is that it is the sole prerogative of the organization to make these decisions based on consideration of the policy, products, services, processes, customer requirements, and any other pertinent factors.

CLAUSE 6.3—PLANNING OF CHANGES

What Does the Clause Say?

When there is a need for changes to the QMS, the changes shall be carried out under controlled conditions (i.e., in a planned and systematic manner) that consider:

- What will be done and why

- Potential adverse consequences

- Maintenance of the integrity of the QMS

- What resources will be required

- Allocation or reallocation of responsibilities and authorities

What Does the Clause Mean?

When and where the organization determines there is a need for changes to the QMS, the changes are required to be carried out

Figure 5.2 Effective change requires careful control.

in a planned and systematic manner (i.e., under controlled conditions). The requirements stated in clause 4.4 for activities that shall be determined for QMS processes also apply to changes to processes and are required to be addressed. If changes are to be effective, careful control is required during the change state (see Figure 5.2).

CONSIDER THE POTENTIAL INTERACTIONS AS THEY APPLY TO YOUR QMS

The planning processes of the QMS interact with most of its other processes. Particular attention should be focused on clauses 4.1, 4.2, 5, 8, 9, and 10.

IMPLEMENTATION TIPS

Uncertainty is a natural phenomenon that cannot be avoided. There is no "opting out" of risk. Risk-based thinking is always needed. But with risk comes opportunity. Failure to deal in a systematic way with known uncertainties of the organization may not just be foolhardy; it also means the organization will have a tendency to avoid situations that could bring great opportunities. The new version of ISO 9001 recognizes this reality.

When the organization is planning the QMS, it is required to determine external and internal issues that are relevant to its purpose (see clause 4.1). Some of the possible considerations for achieving this understanding of external and internal issues are included in notes 1 and 2 of clause 4.1. We recommend a process to recognize changes in this information over time. Our list is similar but broader in scope and is detailed in Chapter 3. This can be—and in many organizations will be—a lot of work! So it is important not to become intimidated or overwhelmed. While the details are the heavy lifting, the concept and its objectives can be illustrated very simply (see Figure 5.3).

The words in the standard related to interested parties are intended to broaden the viewpoint of an organization beyond just meeting the requirements of customers to also consider the requirements of interested parties that are relevant to, and can have an impact on, the organization.

The inclusion of interested parties in the requirements segment of the standard may be either confusing or mystifying (or both) to organizations for several reasons. For example, in the introduction in clause 0.1 the standard states:

This International Standard can be used by internal and external parties, to assess the organization's ability to consistently meet customer, statutory and regulatory

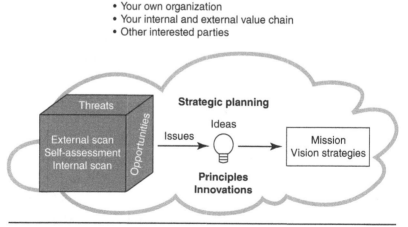

Figure 5.3   Risk assessment as input to planning.

requirements applicable to the products and services it provides, the organization's own requirements and its aim to enhance customer satisfaction.

Also, the scope clause states that the organization needs to demonstrate its ability to consistently provide product or service that meets customer and applicable statutory and regulatory requirements.

Although the front end of ISO 9001:2015 reinforces the historical nature and intent of ISO 9001 as being focused on the customer and customer requirements, clause 4.1 introduces the concept of determining the requirements of interested parties that are relevant and a requirement to monitor and review the information about interested parties.

What this means will vary from organization to organization and from auditor to auditor. Even having a vague idea of the intent of including the concept of interested parties in ISO 9001 and in their QMS, organizations will have difficulty discerning how to approach conformity with this requirement.

What this clause means to us is that the organization shall decide what interested parties, if any, are relevant and shall incorporate into the QMS planning process how it will monitor and review the information about these interested parties and their relevant requirements.

Words like "relevant," "determine," "monitor," and "review" in a requirements standard have the potential to introduce more ambiguity than clarity.

The newness of the concepts and the "vagueness" of the requirements should not cause organizations to ignore or cavalierly address the requirements of clause 6. Integrating the processes required by clause 6 (and by reference clause 4) will strengthen the QMS and the alignment of quality with the strategic and tactical objectives of the organization.

Regarding quality objectives, we recommend the development and deployment of a formal process for addressing this requirement and maintaining and retaining documented information of the objectives that are established as evidence of conformity (i.e., both procedures and records), not because ISO 9001:2015 requires such documentation but rather because it is fundamental to effective management of the organization.

When changes are required to any element of the QMS, it is good practice to consider such changes as similar to the development of a new product and to have a procedure that requires exercising controls such as design review, verification, and validation of proposed changes.

In addition to consideration of interested parties and of risks and opportunities when planning the QMS, organizations should also consider reviewing the QMPs and how they should be addressed. The QMPs will need to be addressed in a unique manner for each organization. They will not all require the same degree of emphasis, depending on the products, services, and customers of the organization. For example, organizations that make medical devices will have priorities that are different from those of a bank or an organization that makes toys. The priorities should be reflected in the processes and controls incorporated into a QMS.

In addition to the QMPs it is also advisable to review the mission and vision statements for the organization. Are they current? Have they been communicated across the organization? Are they understood? The mission, vision, and QMPs are elements of the foundation of the organization (recall Figure 3.2). Effective planning of the QMS requires consistency with the rest of the management system. However, that alone is no longer good enough. The QMS also needs to display seamless integration with all the other elements of the strategic and tactical planning of the organization. Remember that the objective is to create, deploy, and improve a QMS that is embraced by the organization with enthusiasm. The QMS should not be a collection of processes that is tolerated but detached from the objectives of the organization.

Finally, the organization needs to consider how far to go in determining relevant external and internal issues and determining who are interested parties and be able to substantiate its position. Having documented information relevant to its actions in these areas (i.e., procedures and records) will be useful to the organization both to ensure these activities are performed in accordance with the needs of the organization and to avoid ponderous discussions (and potential disagreements) with external auditors.

QUESTIONS TO ASK TO ASSESS CONFORMITY

- Have the risks and opportunities that need to be addressed been identified?

- Have undesired effects been considered?

- Has improvement related to both risks and opportunities and interested parties been considered?

- Have external and internal issues been considered?

- Is there evidence that information about external and internal issues is being monitored and reviewed?

- Have interested parties that are relevant to the QMS been determined?

- Have the requirements of all relevant interested parties been determined?

- Is there evidence that information about interested parties and their relevant requirements is being monitored and reviewed?

- Have quality objectives at relevant functions, levels, and processes been established?

- Is documented information related to quality records maintained?

- Are the objectives consistent with the quality policy?
- Are the objectives measurable?
- Are the objectives monitored?
- Are the objectives updated?
- Are the objectives appropriate?
- Do the objectives address Who? What? When? and How?
- How does the organization ensure that changes to the QMS are carried out in a planned and systematic manner?

DEFINITIONS (REFER TO ISO 9000:2015)

- Documented information
- Function
- Improvement
- Innovation
- Interested party
- Measurement
- Measurement process
- Monitoring
- Objective
- Organization
- Process
- Quality management system
- Quality objective
- Risk

CONSIDERATIONS FOR DOCUMENTED INFORMATION TO BE MAINTAINED AND/OR RETAINED

There is an explicit requirement for documented information in clause 6.2 related to maintaining documented information on the quality objectives.

By inference and reference to clause 4.4, documented information may be required to substantiate that changes to the QMS are carried out in a planned and systematic manner.

In addition, we recommend having documented information (procedures and/or records as appropriate) related to:

- Identification of risks and opportunities

- Consideration of undesired effects

- Consideration of improvement related to both risks and opportunities and interested parties

- Determination of interested parties and their requirements

- Monitoring and review of information about interested parties

6

Clause 7—Support

Innovation is the specific instrument of entrepreneurship. The act that endows resources with a new capacity to create wealth.

—PETER DRUCKER

CLAUSE 7.1—RESOURCES

What Does the Clause Say?

Clause 7.1 has six subclauses:

- 7.1.1 General
- 7.1.2 People
- 7.1.3 Infrastructure
- 7.1.4 Environment for the operation of processes
- 7.1.5 Monitoring and measuring resources
- 7.1.6 Organizational knowledge

Many of the requirements that were in clause 6 of ISO 9001:2008 are now contained in clauses 7.1.1–7.1.4. The requirements for clause 7.1.5 are a recasting of clause 7.6 in ISO 9001:2008. Clause 7.1.6 is a new requirement of ISO 9001:2015.

Clause 7.1.1, *General*, requires the organization to determine and provide the resources needed to establish, implement, maintain, and improve the QMS. It adds a requirement to consider the capabilities of, and constraints on, existing internal resources and the needs of external providers.

Clause 7.1.2, *People*, is an update to the ISO 9001:2008 requirements to provide the human capital necessary for the effective operation of the QMS, including the processes needed. It adds a requirement to meet statutory and regulatory requirements in addition to customer requirements.

Clause 7.1.3, *Infrastructure*, repackages the ISO 9001:2008 requirements, taking the examples that were in the body of ISO 9001:2008 and making them a note, and so there is no substantive difference from the past for this requirement.

Clause 7.1.4, *Environment for the operation of processes*, updates the previous requirements for work environment with an update to the note to suggest consideration of social and psychological environments, ergonomics, and cleanliness while dropping the previous examples of noise, lighting, and weather as areas to be considered. The note implies a change in emphasis and intent.

The requirements for monitoring and measuring resources are incorporated in clause 7.1.5, which has two subclauses: 7.1.5.1, *General*, and 7.1.5.2, *Measurement traceability*. In ISO 9001:2008 these requirements were in clause 7.6. The requirements are similar to those of 2008 and can be summarized as follows:

- Determine the resources needed for valid and reliable monitoring and measuring results

- Ensure the resources provided are suitable

- Ensure the resources provided are maintained

- Retain appropriate documented information as evidence of fitness for purpose

- Where measurement traceability is a requirement or considered necessary by the organization, measuring instruments shall be verified or calibrated at specified intervals or prior to use against measurement standards traceable to international or national measurement standards

• The basis used for calibration or verification shall be retained as documented information if no standard exists

• Measuring instruments shall be identified

• Measuring instruments shall be safeguarded from adjustments, damage, or deterioration

• When an instrument is found to be defective, determine whether there is any adverse effect on the validity of previous measurement results

Old-school metrologists will be either amused or annoyed by a few of the somewhat subtle changes that have been made, including replacing the word "equipment" with "resources" and "calibrated equipment" with "instruments."

A new requirement is incorporated into ISO 9001:2015 clause 7.1.6, *Organizational knowledge*. The requirement, while somewhat vague and subject to a wide spectrum of understanding (and hence a potential for many issues related to auditability), can be summarized as follows:

• Determine the knowledge necessary for the operation of processes

• Maintain necessary knowledge and make it available, as appropriate

• Consider the organization's current knowledge inventory and determine how to acquire or access the necessary additional knowledge to address changing needs and trends

Additional information related to this requirement can be found in Annex A (A.7, *Organizational knowledge*).

What Does the Clause Mean?

The six subclauses of clause 7.1 address the various resources needed to establish, implement, maintain, and improve the QMS. In addition to clause 7.1 in ISO 9001:2015, clauses 8.3 and 9.3 also refer to resources.

Clauses 7.1.1–7.1.4 contain some minor changes from the prior edition of ISO 9001, but the content and intent are similar and the requirements are clear, understandable, and not subject to misinterpretation.

An organization that was in compliance with the ISO 9001:2008 requirements will have little, if any, difficulty tweaking processes and existing documented information to ensure compliance with the requirements of these clauses.

Clause 7.1.5 and its two subclauses, 7.1.5.1 and 7.1.5.2, "repackage" requirements similar to those in the prior edition of ISO 9001. Those involved in metrology processes will notice a change in terminology: the word "equipment" has been replaced with "resources," and "calibrated equipment" has been replaced with "instruments."

It is interesting to us that confirmation of software that is integral to much contemporary instrumentation is not stated as a requirement. We believe that the intent of the words "ensure the resources provided are suitable" requires organizations to include determination that software that is an element of a measuring or monitoring system functions in accord with requirements (both doing what it is supposed to do and not doing what it is not intended to do). Ensuring software integrity is an implicit requirement, and the functionality of software should be incorporated into metrology control processes.

The requirements of clause 7.1.6 include words like "determine" and "to the extent necessary" and "consider." Although the requirements may seem vague, they are important. We believe what this clause means is that the organization should have a process for periodically assessing what knowledge is required for both the short term and the long term and decide on actions to either acquire, retain, or defer action related to the required or desired knowledge.

Consider the Potential Interactions as They Apply to Your QMS

The following are primary areas where clause 7.1 can potentially interact with other clauses of ISO 9001:2015:

* Clause 5 Leadership

* Clause 6 Planning

* Clause 8 Operation

* Clause 9 Performance evaluation

* Clause 10 Improvement

Implementation Tips

Little updating effort, if any, will be needed to ensure conformity to the requirements of clause 7.1, with the exception of clause 7.1.6, which is a new requirement to be addressed.

We recommend review of the language in the new standard and then tweaking processes and documented information, as appropriate. These changes could include incorporation of any additions, deletions, or modifications that are perceived as necessary or desirable to at least conform to the requirements. But there is opportunity to take the time to also improve process effectiveness and efficiency (even though efficiency is not an explicit element of ISO 9001). The issuance of ISO 9001:2015 is a unique opportunity to consider potential improvements even if current processes are already in conformity.

It may be helpful to review Annex A—*Clarification of new structure, terminology, and concepts*—since this annex contains guidance that may enhance understanding of the standard and facilitate creation or improvement of processes. Remember, however, that Annex A is informative, not normative.

Questions to Ask to Assess Conformity

- How are required resources determined?

- What needs to be obtained from external providers?

- How are infrastructure requirements determined and assessed for adequacy?

- How are environmental requirements necessary for the operation of the organization's processes and to achieve conformity of products and services determined and assessed for adequacy?

- Has the organization identified the measurements to be made?

- Has the organization retained appropriate documented information as evidence of fitness for purpose of monitoring and measurement resources?

- Are monitoring and measurement devices calibrated and adjusted periodically or before use against devices traceable to international or national standards?

- Is the basis used for calibration recorded when traceability to international or national standards cannot be done, since no standards exist?

- Are monitoring and measurement devices protected from damage and deterioration during handling, maintenance, and storage?

- How is determination of knowledge necessary for the operation of the organization's processes obtained?

• If changing needs and trends are anticipated, how does the organization consider how to acquire or access the necessary additional knowledge?

CLAUSE 7.2—COMPETENCE

What Does the Clause Say?

The requirements addressed in clause 7.2 can be summarized as follows:

• Determine the competence required for person(s) doing work that affects quality performance

• Ensure competence on the basis of appropriate education, training, or experience

• Take actions to acquire necessary competence, and evaluate effectiveness of actions taken

• Retain appropriate documented information as evidence of competence

The note to this clause provides examples of the kinds of activities that can be pursued to obtain or enhance competence (e.g., the provision of training, mentoring, the reassignment of persons, or the hiring or contracting of competent persons).

What Does the Clause Mean?

The requirements are similar to those in clause 6.2.2 of ISO 9001:2008. The requirements mean that the organization shall know what skills, experience, education, and training are necessary to perform work to meet requirements. Where there is a disparity between the skills required and the capability of personnel, the organization is required to take action to eliminate the disparity. There is also a requirement that records be retained to provide objective evidence of competence of personnel.

Consider the Potential Interactions as They Apply to Your QMS

The following are primary areas where clause 7.2 can potentially interact with other clauses of ISO 9001:2015:

- Clause 6 Planning

- Clause 8 Operation

- Clause 9 Performance evaluation

- Clause 10 Improvement

Implementation Tips

We recommend that organizations give these requirements serious consideration. This is important because failure cost data (e.g., from a cost of quality process) often indicate that personnel-related nonconformity has impact on both internal operations and customer satisfaction. While this is obvious in the service sector, we have also found it to be true in construction and manufacturing. The issue of competence is compounded by actions in HR, which often views its responsibility as ensuring that personnel are hired and placed in positions, and not as ensuring that *competent* personnel are hired and placed in positions. Top management can also reinforce the requirement that competent personnel are assigned to work even when pressure to complete a project, deliver a service on time, or ship a product arises (see also clause 5.1.1).

There are several ways to address the requirement to take actions to acquire the required competence where there is a disparity between the required skills and the capability of personnel. The typical reaction is to "train the people," but that is seldom the only action that is needed or should even be taken. Management should consider the alternatives. Perhaps the process itself or related equipment is not capable of meeting requirements. If the process is flawed, training the people may well just drive down morale. Rather, the process may require improvement.

Perhaps the time allotted to do required activities or the arrangement of the facilities is insufficient. The people may know how to do the job, but "the system" impedes their performance and requires improvement. Transfer of personnel is another option that may be required. We recommend management not jump to an easy resolution until confidence has been established that the actions to be taken will improve the system.

Processes across the organization for hiring competent personnel and maintenance of competence for all activities are the intent of the requirements of clause 7.2. Such processes are also integral elements of organizational sustainability.

Questions to Ask to Assess Conformity

- How does the organization determine the competence necessary for person(s) doing work?

- How does the organization ensure that these persons are competent on the basis of appropriate education, training, or experience?

- Is appropriate documented information available as evidence of competence?

- Where action is taken to acquire necessary competence, is the effectiveness of the actions taken evaluated?

CLAUSE 7.3—AWARENESS

What Does the Clause Say?

Clause 7.3 requires individuals doing work to have awareness of:

- The quality policy

- Relevant quality objectives

- The contribution they provide to the effectiveness of the system, including improved quality performance

- The implications of not conforming with requirements

What Does the Clause Mean?

The scope of the awareness requirements clause has been expanded to include awareness of policy and objectives and of the implications of not meeting requirements. Alignment across the organization of policy, objectives, and engagement of people requires an awareness of organizational policy and objectives, and the creation of this awareness is what this clause means. To ensure the required awareness is created and maintained, the organization needs to deploy processes to ensure consistent communication of the required information.

Consider the Potential Interactions as They Apply to Your QMS

The following are primary areas where clause 7.3 can potentially interact with other clauses of ISO 9001:2015:

- Clause 6 Planning
- Clause 8 Operation
- Clause 9 Performance evaluation

Implementation Tips

Although no documented information is required, we recommend a process be deployed to ensure consistent conformity to the requirements indicated earlier.

It is not sufficient to post the quality policy in the lobby of an organization or to instruct managers to inform the workers of the objectives or to discuss the role of each worker in an annual review.

Organizations have devised numerous mechanisms for meeting the intent of this clause. Examples of ideas to create awareness of policy and objectives and the importance of each employee are:

- E-mails to all employees from top management
- Personal interaction with employees by top management

- Meetings hosted by top management to inform employees of the "state of the organization"

- Postings on bulletin boards

- Letters to all employees

It also is appropriate to use QMPs 2 and 3, on leadership and engagement of people, since the QMPs are foundational to ISO 9001 content.

Also, the organization should consider formalizing its processes for communication and interaction to ensure information is available, understood, and useful. Such a process could address who communicates what, and when and how.

Questions to Ask to Assess Conformity

- Are all members of the organization aware of the quality policy and its meaning?

- Are all members of the organization aware of their own objectives and those of the organization?

- How is the importance of the contribution of each individual in the organization to the effectiveness of the QMS communicated?

CLAUSE 7.4—COMMUNICATION

What Does the Clause Say?

A short but important clause, clause 7.4 requires the organization to determine the internal and external communications relevant to the QMS, including:

- What it will communicate

- When to communicate

- With whom to communicate

- How to communicate

What Does the Clause Mean?

These requirements are similar in intent to previous requirements, but they are more explicit and a little more expansive since external communication is now included. They provide structure on communication to ensure it occurs at least to a minimal degree and that content is consistent with organizational needs. Timing of communications is often critical and should be given careful consideration. Absent a requirement, such activities can be viewed as less urgent than day-to-day activities and may be marginalized or ignored.

Consider the Potential Interactions as They Apply to Your QMS

The following are primary areas where clause 7.4 can potentially interact with other clauses of ISO 9001:2015:

- Clause 5 Leadership
- Clause 6 Planning

Implementation Tips

In today's organizations, great communication is essential. It is also more critical than ever for all aspects of the QMS to perform in an effective manner every day. Rapid and varied communication mechanisms are needed so all members of the organization receive regular and accurate feedback on performance. The feedback needs to be timed so information is actionable, meaningful, and timely. Many organizations are evolving ways to use social media and other advanced mechanisms to enhance communications and engagement of all members of the organization.

We recommend a formal process be deployed that indicates who will do what and how often.

Questions to Ask to Assess Conformity

- How has the organization determined its internal and external communications channels?

- Is responsibility clear for who will communicate what and when and how?

CLAUSE 7.5—DOCUMENTED INFORMATION

What Does the Clause Say?

Clause 7.5 has three subclauses: 7.5.1, 7.5.2, and 7.5.3 (with 7.5.3 having two untitled subclauses, 7.5.3.1 and 7.5.3.2). The clauses are titled:

- 7.5.1 General

- 7.5.2 Creating and updating

- 7.5.3 Control of documented information

The most significant change in 2015 is in the nomenclature. The previous terms of "documents" and "records" have been replaced with "documented information." Since this change in terminology can have a significant impact on the overall documentation of the QMS (i.e., procedures and records), let us understand the ISO 9000:2015 definition of documented information—"information required to be controlled and maintained by an organization."

The essence of the requirements for documented information is that the QMS shall include:

- All documented information required by ISO 9001:2015

- Documented information that the organization determines as necessary for effective operations

When documented information is created and updated, the organization shall ensure appropriate:

- Identification and description (e.g., a title, date, author, or reference number)

- Format (e.g., language, software version, graphics) and media (e.g., paper, electronic)

- Review and approval for suitability and adequacy

Documented information required by the QMS and by this International Standard shall be controlled to ensure:

- Availability and suitability for use, where and when they are needed

- Adequate protection

For the control of documented information, the organization shall address the following activities, as applicable:

- Definition and maintenance of distribution, access, retrieval, and use

- Storage and preservation

- Control of changes

- Retention and disposition are maintained

- Documents of external origin (such as industry and customer specifications and standards) are identified and controlled

What Does the Clause Mean?

The requirements for documented information are explicit and clear, and the intent is consistent with ISO 9001:2008 requirements for documented procedures and records. Although the requirements for documented information included in clause 7.5 and many other places may appear to be diffused, the organization is still required to ensure that documented information is controlled in accordance with at least the specific activities included in clauses 7.5.1, 7.5.2, and 7.5.3.

The organization should, in particular, consider its approach to ensuring conformity with the requirement stated in clause 7.5.1b, which states that the organization's QMS shall include "documented information determined by the organization as being necessary for the effectiveness of the quality management system." *This means the organization shall determine what documented information is required to conform to customer, regulatory, and internal requirements and ensure control of the availability, suitability, and maintenance of that documented information. Proper understanding and application of these concepts and their interaction with the process approach is, in our opinion, one of the most important core concepts for successful QMS implementation.*

The standard also requires that controls be implemented for documented information that arises from outside the organization (e.g., industry standards, customer specifications, international and national standards) and for the retention and disposal of documented information.

Consider the Potential Interactions as They Apply to Your QMS

The requirements of clause 7.5 interact with all other elements of ISO 9001. Availability and maintenance of procedures and records (to use the "old" terminology) are fundamental to planning and operating all the processes of the QMS under controlled conditions (see clause 8.1). Correct application of the interaction with the process approach is crucial to success.

Implementation Tips

We offer a few general comments regarding documented information: (1) if an organization desires to continue to use documents and records nomenclature, this is permissible and may be required in some sectors; (2) documented information is expected to vary from organization to organization (see note to clause 7.5.1); (3) Annex A.6 contains additional guidance on documented information.

To avoid confusion in the organization between documents (which can be changed) and records (which can be corrected but cannot be changed), we recommend organizations give careful consideration to maintaining a clear differentiation between documents and records even though the standard does not require organizations to make such a distinction. It is important to understand the intended message of the new wording. Where the term "maintain documented information" is used, the intent is that the documented information be kept up to date, much as a documented procedure would be, using the older terminology. Where the term "retain documented information" is used, the intent is similar to the old requirement for a record. See Table 6.1.

We also recommend development of written procedures addressing the requirements of clause 7 in addition to all the requirements of the organization. Particular attention to records is justified for many reasons, including resolution of customer issues, resolution of "disputes" with outside auditors, and the capability of "proving" the organization is exercising prudent

Table 6.1   Documented information (2015) versus documents and records (2008).

Old term	New term	Purpose of documented information	Changes and corrections
Documented procedure	Maintain documented information	Provide up-to-date information for process operations	Keep up to date with changes
Record	Retain documented information	Provide objective evidence of activities conducted	No changes but corrections are permitted

judgment in the determination of the conformity of products and services to requirements.

It may be helpful to review the informative guidance in A.1 and A.6 in the annex, which address documented information. Annex A is not normative (i.e., it does not contain requirements), but it does provide information that can facilitate effective achievement of conformity.

The new terminology of "documented information" that replaces "documented procedures" and "records" in the previous editions of ISO 9001 may be confusing—and even dangerous—for some organizations to adopt (e.g., those with regulatory requirements). We recommend careful consideration before making any changes to achieve consistency with the new nomenclature. In a number of industries, the distinction between documented procedures and records is ingrained, so continued use of the historic terminology of documented procedures and records may very well be prudent. In other words, do not change unless your organization has a good reason to change. For most organizations we see **NO** value in rushing to embrace the new language for internal use.

Users of the standard should also be aware that, although the primary requirements for documented information are in clause 7.5, documented information is also addressed in at least 21 other places in clauses 4–10. See Table 6.2 for examples of clauses that mention "documented information."

In the table:

- "M" indicates a requirement to maintain documented information related to a requirement. In earlier versions of ISO 9001 this would have been expressed as a requirement for a documented procedure(s).

- "R" indicates a requirement to retain documented information related to a requirement. In earlier versions of ISO 9001 this would have been expressed as a requirement for records.

Table 6.2   References to documented information.

4.3 M	8.1 D & R	8.5.2 R
4.4.2 M	8.2.3.2 R	8.5.3 R
5.2.2 M	8.3.2 D	8.5.6 R
6.2.1 M	8.3.3 R	8.6 R
7.1.5.1 R	8.3.4 R	8.7.2 R
7.1.5.2 R	8.3.6 R	9.1.1 R
7.1.6 A	8.4.1 R	9.2.2 R
7.2 R	8.5.1 D & A	9.3.3 R
7.5.3.1 A	8.3.6 R	10.2.2 R
7.5.3.2 R		

- "D" indicates a requirement for the organization to determine the documented information needed to achieve effective conformity with a requirement.

- "A" indicates that documented information shall be available.

Since ISO 9001:2015 will be in play through 2020 (or longer), we recommend not only considering the current requirements but also building in flexibility to facilitate adoption of innovative concepts and adaption to changing circumstances.

Questions to Ask to Assess Conformity

- Has the documented information required by this standard been established?

- Is all documented information determined by the organization as being necessary for the effectiveness of the QMS available?

- Is documented information approved for adequacy prior to use?

- Is documented information reviewed and updated as necessary?

- Are changes to documented information reapproved to ensure adequacy prior to use?

- Is the current revision status of documented information maintained?

- Are relevant versions of applicable documented information available at points of use?

- Is there a process to ensure that documented information remains legible, identifiable, available when and where required, and retrievable?

- Is documented information of external origin identified?

- Is obsolete documented information that is retained for any purpose suitably identified to prevent unintended use?

- Is documented information protected (e.g., from loss of confidentiality, improper use, or loss of integrity)?

DEFINITIONS (REFER TO ISO 9000:2015)

- Change control

- Competence

- Conformity

- Determination

- Documented information

- External supplier

- Improvement

- Infrastructure

- Innovation
- Knowledge
- Measurement
- Monitoring
- Objective
- Organization
- Performance
- Process
- Product
- Quality management system
- Quality policy
- Regulatory requirement
- Requirement
- Service
- Statutory requirement
- Traceability

CONSIDERATIONS FOR DOCUMENTED INFORMATION TO BE MAINTAINED AND/OR RETAINED

- Documented information is required as evidence of fitness for purpose of monitoring and measurement resources

- Documented information as evidence of competence is required

- All documented information required by the ISO 9001: 2015 standard is required

- All documented information determined by the organization to be necessary for the effectiveness of the QMS is required

- We recommend the organization consider development or modification of existing documented information (i.e., procedures and records), as appropriate, to incorporate any additions, deletions, or modifications that are perceived as necessary or desirable to conform to the requirements of clauses 7.1.1–7.1.4, and that those changes be made with a careful eye on attaining improved process effectiveness and efficiency

- For clause 7.1.5 we recommend that documented information (i.e., records and procedures) exist to provide clear evidence of conformity with the ISO 9001 requirements as well as with internal requirements

- For clause 7.1.6, *Organizational knowledge*, since the requirements are somewhat vague and subject to a wide spectrum of understanding (and hence a potential area for many issues related to auditability), we recommend the development (or modification) of documented information (i.e., a procedure) that states what actions shall be taken and what records shall be maintained

- We recommend development, deployment, and maintenance of documented information to ensure the requirements are met on an ongoing basis for creating awareness of the following:

  — Quality policy

  — Quality objectives

  — Importance of the contribution of every employee of the organization

- We recommend a formal process be deployed that stipulates communication requirements, including, at the very least, who will do what and how often

7

Clause 8—Operation

Being busy does not always mean real work. The object of all work is production or accomplishment and to either of these ends there must be forethought, system, planning, intelligence, and honest purpose, as well as perspiration. Seeming to do is not doing.

—Thomas A. Edison

INTRODUCTION

Clause 8 contains much of what was in ISO 9001:2008 clause 7, *Product realization*. Clause 8 has seven subclauses, four of which have their own subclauses.

This chapter addresses the requirements for clause 8 in two parts. Part 1 covers front-end processes such as planning, design, and externally provided processes, products, and services. Part 2 covers the requirements related to production and service provision.

The top portion of Figure 7.1 provides a model for Part 1 of clause 8 subclauses and their interaction.

The specifics of the clauses addressed in Part 1 are as follows:

- 8.1 Operational planning and control

- 8.2 Requirements for products and services

  — 8.2.1 Customer communication

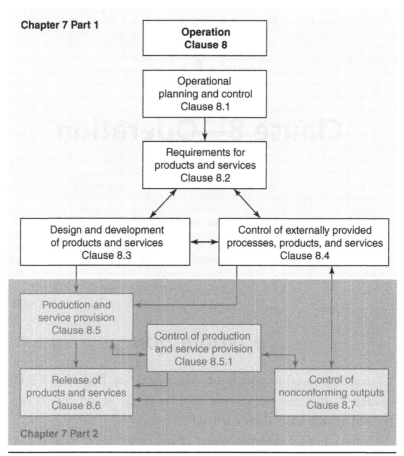

Figure 7.1 Operational processes for planning, through external provision of products and services.

—8.2.2 Determination of requirements related to products and services

— 8.2.3 Review of requirements related to products and services

— 8.2.4 Changes to requirements for products and services

• 8.3 Design and development of products and services

— 8.3.1 General

— 8.3.2 Design and development planning

- 8.3.3 Design and development inputs
- 8.3.4 Design and development controls
- 8.3.5 Design and development outputs
- 8.3.6 Design and development changes
- 8.4 Control of externally provided processes, products, and services
  - 8.4.1 General
  - 8.4.2 Type and extent of control
  - 8.4.3 Information for external providers

These clauses cover planning of the operational aspects of the system. Every organization must get these up-front activities right, or the end product or service is unlikely to meet customer needs and requirements. These four clauses have interactions that can be critical to the overall business. For example, the design and development process is often directly involved in reviewing product requirements from the customer. And it is normal for sourcing and procurement decisions to be made during the review of quotations and preparation of proposals. This means controls for externally provided products are, or at least should be, decided on at this early planning stage.

The specifics of the clauses addressed in Part 2 of this chapter are as follows:

- 8.5 Production and service provision
  - 8.5.1 Control of production and service provision
  - 8.5.2 Identification and traceability
  - 8.5.3 Property belonging to customers or external providers
  - 8.5.4 Preservation
  - 8.5.5 Post-delivery activities
  - 8.5.6 Control of changes
- 8.6 Release of products and services
- 8.7 Control of nonconforming outputs

The clauses covered in Part 2 involve the actual production of the product or delivery of the service.

PART 1: CLAUSES 8.1–8.4—OPERATIONAL PLANNING AND CONTROL THROUGH CONTROL OF EXTERNALLY PROVIDED PROCESSES, PRODUCTS, AND SERVICES

Clause 8.1—Operational Planning and Control

What Does the Clause Say?

Clause 8.1 requires that operations be conducted through processes that are planned and controlled regardless of whether the organization or an outside party performs the process. Requirements for products and services are required to be determined and criteria established for acceptance. Identification of resources needed to achieve conformity is required. Planned changes are required to be controlled and action taken to mitigate the effects of unintended consequences of changes. Documented information is required to be kept (retained) to demonstrate conformity of product and service to requirements and that processes have been carried out as planned.

What Does the Clause Mean?

While the topic of clause 8.1, *Operational planning and control*, is similar to the title of ISO 9001:2008 clause 7.1, *Planning of product realization*, the actual requirements have quite a different focus. Where the individual planning step in the 2008 version focuses on determining how to verify conformity, the 2015 version is oriented around the notion of managing and adequately resourcing a set of processes so that a state of control is achieved even when intended or unintended changes occur. It also indicates a requirement to review consequences and mitigate adverse effects as necessary. The clause also requires that the organization keep documented information to have confidence that these processes have been carried out as required. Note that the requirement for processes to accomplish all the operational activities is not repeated in each clause. A lack of repetitious requirements in each clause does not mean processes and documented information are not required.

The first sentence of clause 8.1 makes the point that *the organization "shall plan, implement and control the processes needed to meet the requirements for the provision of products and services."* It also reinforces the relationship between clauses 4.4 and 6. Therefore, even though ISO 9001:2015 may appear to some to have reduced the requirements for processes and controls, we believe clause 8.1, coupled with clauses 4 and 6, requires an organization to define, document, control, and keep records at least at the same level as previously required, and perhaps to an even greater level of comprehension.

Consider the Potential Interactions as They Apply to Your QMS

- Clause 4 Context of the organization that drives changes to products and processes

- Clause 6.1 Actions to address risks and opportunities as appropriate

- Clause 7.1 Resources including measurement processes, infrastructure, operational environment, people, and so on

- Clause 8.4 Control of externally provided processes, products, and services

- Clause 9 Performance evaluation

- Clause 10 Improvement

Implementation Tips

Many organizations long ago adopted the process approach to managing operations, and thus may already be close to conforming. We recommend review of the language in the new standard and tweaking processes and documented information, as appropriate. The organization needs to incorporate any additions, deletions, or modifications that are perceived as necessary or desirable to conform with these more process-oriented requirements and possibly to improve process effectiveness. Subtle "new" requirements related to control of changes and mitigating adverse effects should also be considered. This is another example of the 2015 edition providing a unique opportunity to consider potential improvements even if processes are compliant as they are currently implemented.

On the other hand, some organizations may not have embraced the process approach to operational controls. For these organizations, the planning activity required by clause 8.1 may not be a trivial exercise. Less documentation may be required, but ISO 9001:2015 requires that the organization understand the processes needed to deliver conforming products to customers. These processes must be understood not only with respect to the products themselves but also in the broader context of the objectives of the organization and any other requirements of the QMS (including interested parties and risks and opportunities).

It may be advisable to:

- Create a quality plan for a product or service to describe how the QMS will be modified and applied to all operations. Such a plan could include or reference procedures and records to be maintained and analyzed.

- Consider using the product design and development process approach for designing processes. This is a requirement in the automotive industry. It has become a best practice demonstrated in many organizations even though ISO 9001 does not explicitly require adherence to the design and development requirements for internal process designs. We believe such discipline enhances both the effectiveness and the efficiency of processes.

- Identify key performance measures for both products and processes and align them with your quality and business objectives.

Questions to Ask to Assess Conformity

- Is there evidence of planning of production processes?

- Does the planning extend beyond production processes to encompass all operational processes?

- Are processes planned, controlled, and operated as outlined in clause 4.4?

- Is the planning consistent with other elements of the QMS?

- Are the operational documented information and planning output adequate for the organization's needs?

- Are support needs and resources defined during the planning process, and do they appear to be adequate?

- Does the planning define the documented information that must be retained to provide confidence in the conformity of the processes and resulting product?

Definitions (Refer to ISO 9000:2015)

- Change control

- Configuration management

- Design and development

- Documented information

- Measurement

- Measurement management system

- Measurement process

- Measuring equipment

- Process

- Product

- Quality plan

- Quality planning

- Requirement

- Service

- Specification

Considerations for Documented Information to Be Maintained and/or Retained

- The only requirement in clause 8.1 for documented information is to retain or maintain documented information as necessary to provide confidence that the processes have been carried out as required

- Organizations should also consider maintaining documented information describing the operational processes and how they are to be carried out

- The requirement to plan, implement, and control the processes needed to meet the requirements for the provision of products and services would be very difficult to achieve if documented information is not created and maintained for all processes of the QMS

Clause 8.2—Requirements for Products and Services

Clause 8.2 has four subclauses:

8.2.1 Customer communication

8.2.2 Determination of requirements related to products and services

8.2.3 Review of requirements related to products and services

8.2.4 Changes to requirements for products and services

Clause 8.2.1—Customer Communication

What Does the Clause Say?

Clause 8.2.1 requires the organization to conduct communications with customers. The detail requirements for communications with customers include:

- Providing product- and service-related information

- Handling customer orders of all types and changes thereto

- Getting customer feedback including complaints

- Exercising appropriate controls for any customer-owned property

- Establishing requirements for contingency actions

What Does the Clause Mean?

Clause 8.2.1 requires processes to accomplish specific types of information exchange. Similar information exchange requirements are in ISO 9001:2008 clause 7.2.3, which is also titled *Customer communication*. Both versions require three specific types of communication with customers to be included in the organization's processes:

- Product and service information, including customer requirements

- Documented agreements with the customer, such as contracts, orders, changes, and other information needed to meet customer requirements

- Customer feedback, including complaints

Clause 8.2.1 of ISO 9001:2015 also has two additional requirements. It requires the organization to include control of:

- The handling and treatment of customer-owned items was covered in great detail in clause 7.5.4 in ISO 9001:2008.

The specific requirements of the 2008 version have been significantly simplified.

- Any contingency actions that are relevant.

Consider the Potential Interactions as They Apply to Your QMS

- Clause 4.2b Relevant requirements of interested parties

- Clause 4.4 Quality management system and its processes

- Clause 5.2 Policy

- Clauses 6.2, 7.3, 9.3.2 Quality objectives

- Clause 7.1.6 Organizational knowledge

- Clause 7.3 Awareness

- Clause 7.4 Communication

- Clause 8.1 Operational planning and control

- Clause 8.2.1 Customer communication

- Clause 8.2.3 Review of requirements for products and services

- Clause 8.5.3 Property belonging to customers or external providers

Implementation Tips

Clause 8.2.1 is similar to clause 7.2.3 of ISO 9001:2008. The point of grouping these items under customer communications is to emphasize that these communications need to be systematically planned like all other processes. In doing so, consider the information in clause 7.4 on communication and the requirements related to process management in clause 4.4. We recommend organizations adopt the philosophy of QMP 7, *Relationship management*, in developing communication processes for customers. If the customer is the organization's most import contact, shouldn't we concentrate some key planning effort on the processes used to communicate with them?

The reader is also referred to the introduction discussed in Chapter 1. It emphasizes the role of the organization to meet customer, statutory, and regulatory requirements applicable to its products and services and its aim to enhance customer satisfaction.

Questions to Ask to Assess Conformity

- Is there evidence of planning of communications with customers?

- Are there processes that cover each type of customer communication that applies in the organization's circumstances?

Definitions (Refer to ISO 9000:2015)

- Conformity

- Conformity contract

- Contract

- Customer

- Customer satisfaction

Considerations for Documented Information to Be Maintained and/or Retained

It is generally necessary and we recommend that a careful record be maintained of the requirements. Often the process involves multiple discussions, reviews (clause 8.2.2), and even early design and development work (clause 8.3). Carefully thought-out methods are needed to efficiently retain this input information for later use in the design process and as input to resolution of disputes that may arise.

Clause 8.2.2—Determination of Requirements for Products and Services

What Does the Clause Say?

Clause 8.2.2 requires the organization to determine the requirements related to its products and services. This includes:

- Establishing a process for determining the requirements for the products offered to potential customers

- Determining requirements of the customer

- Determining requirements for the organization

- Determining requirements from applicable statutes and regulations

- Determining that the organization has the ability to meet the requirements and substantiate claims related to its products and services

What Does the Clause Mean?

One of the key things that the communication with customers needs to ensure is that customer requirements and other requirements for the product or service are clearly understood. But communication and understanding of customer requirements is only one piece of the requirements puzzle. Many products are regulated and customers may have no knowledge of the regulatory or statutory details. Often the organization has learned key things that must be done a certain way for the product or service to meet customer requirements. Customers cannot be expected to know about many of these things. It is the organization's responsibility to understand all these requirements and their specific application. It is also the organization's responsibility to determine whether it can successfully deliver conforming product or service to the customer.

Consider the Potential Interactions as They Apply to Your QMS

- Clause 4.4 Quality management system and its processes

- Clause 5.2 Policy

- Clause 7.5 Documented information

- Clause 8.2.1 Customer communication

- Clause 8.2.3 Review of requirements for products and services

Implementation Tips

Conformity is not difficult for organizations providing off-the-shelf catalog products manufactured to published specifications or standardized services with normal delivery requirements. However, if customers are purchasing complex systems with custom engineering and software according to a complex set of commercial terms, it is essential to obtain a clear understanding of customer requirements by whatever means possible, including activities such as holding face-to-face meetings and attending pre-bid meetings.

Full determination of customer requirements can be an iterative process. Often there are known issues that may evolve into real requirements at a later stage. In such cases, documentation of the open issues and providing for the attendant business risk may prove to be an acceptable approach to meeting the requirements of this clause.

The determination of customer requirements is a critical activity and generally involves several functions and levels in an organization. It is recommended that you maintain documented information to describe the process for determination of all aspects of product and service requirements. The documented information should include both product requirements specified by the customer and product requirements not specified by the customer but necessary for intended or specified use. Also, unique regulatory and statutory requirements should be considered as well as commercial terms and conditions.

Clause 8.2.2b requires that the organization have the ability to meet requirements. Often with advanced products there is a need to advance the state of the art as product development progresses. Such situations should be clearly identified and the business risks understood. In such cases, the defined requirements could be the development of the needed technological advance.

To avoid customer complaints or dissatisfaction, even for "requirements" that are not clearly stated (e.g., regulatory requirements or marketplace norms), we recommend that the organization consider a comprehensive understanding of customer requirements, perhaps even performing a failure modes and effects analysis (FMEA) on the processes as a form of risk assessment.

Questions to Ask to Assess Conformity

- Is there a holistic process for determining requirements, including customer requirements, regulatory and statutory requirements, commercial terms and conditions, and the organization's own requirements?

- Is there a process for determining the organization's capability to meet requirements in specific applications?

- Is there a process for substantiating the validity of the organization's claims for the products and services it offers?

Definitions (Refer to ISO 9000:2015)

- Knowledge

- Management system

- Product

- Quality characteristic

- Requirement

Considerations for Documented Information to Be Maintained and/or Retained

- Clause 8.2 and its three subclauses have no specific requirement for maintaining documented information. Clause 7.5.1 requires the organization to determine the documented information to be maintained and retained.

- Since the review process required in clause 8.2.2 is often iterative, retention of documented information of review results (e.g., who reviewed what, when, and using what method) can be a practical necessity. While clause 8.2.2 does not require retention of any documented information on these determinations, it is recommended to have such records.

Clause 8.2.3—Review of Requirements for Products and Services

What Does the Clause Say?

Clause 8.2.3 states the obligation of the organization to review the requirements of products and services, which includes:

- Customer-specified requirements for the product or service, including any requirements for delivery or post-delivery actions

- Requirements known to be needed by the organization even though not specified by the customer

- Applicable statutory and regulatory requirements applying to the product or service

- Requirements of the final contract or order differing from those previously provided by or discussed with the customer

The review is required to:

- Be performed prior to the organization's commitment to produce the product or service

- Ensure resolution of all order requirements that may differ from those previously defined

- Include confirmation of the requirements in cases where the customer does not provide documented requirements

- Retain documented information on the results of the review

What Does the Clause Mean?

The requirements of this clause are similar to those of clause 7.2.2 in ISO 9001:2008.

The acceptance of an order or the submission of a quote or tender by an organization obliges the organization to meet the conditions stated in the order or to provide the goods and services included in the scope of the quotation or tender. The obligation assumed by the organization includes not only the products defined but also ancillary items such as conformance to stated delivery dates, adherence to referenced external standards, and compliance with the commercial terms and conditions applicable to the order, contract, quote, or tender as well as applicable statutory and regulatory requirements.

The complexity of the order/quote review process depends on the products and services of the organization. A process for reviewing oral orders for off-the-shelf products with 24-hour delivery (e.g., software packages) will differ considerably from a process for reviewing a large order for a one-of-a-kind product with a two-year delivery (e.g., an order for a control system for an electric power-generating station). The review process must also accommodate, as applicable, electronic orders, blanket orders with periodic releases, unsolicited orders, orders through distributors or representatives, faxed orders, and an almost infinite combination of these and other possibilities. If the organization is involved in internet sales, creative thinking will be required to efficiently review customer requirements.

Consider the Potential Interactions as They Apply to Your QMS

• Clause 4.4 Quality management system and its processes

• Clause 5.2 Policy

• Clause 7.5 Documented information

• Clause 8.2.1 Customer communication

- Clause 8.2.2 Determination of requirements related to products and services

- Clause 8.3 Design and development of products and services

Implementation Tips

With such a spectrum of possibilities, what is an organization expected to do to conform to the requirements? The first step should be to develop a clear understanding of the nature of the various kinds of customer requirements and fully understand each communication channel involved. If, for example, an organization publishes a catalog and accepts only written orders for catalog-listed items to standard delivery times, then the order or contract-review procedure can be simple. The process could be a designated individual (e.g., a manager, a clerk) reviewing, initialing, and dating the written order. This simple process can be used as valid evidence that requirements can be met. If an organization must address possibilities that occur only rarely, the organization could simply note in documented information (i.e., a procedure) that any circumstances different from standard terms and conditions will be addressed by a specific quality plan. Such a plan can be generated as the unique occasion arises.

Thus, a simple order-entry process can have a very simple, brief, and effective contract-review process. For the large, complex contracts or quotations, the review process may involve many organizational entities such as engineering, manufacturing, legal, finance, and quality assurance. Accordingly, the procedures governing such reviews can be complex and lengthy.

A good guideline to keep in mind when developing a process to address the specific requirements of clause 8.2.3 is to *balance the risks to the organization with the effort expended* in a review of customer requirements, keeping in mind that the purpose of the review is to add value and not to create a bureaucratic morass. We recommend that a formal process be deployed that indicates who will do what and how often.

Questions to Ask to Assess Conformity

- Does a process exist for the review of identified customer requirements before commitment to supply a product to the customer?

- Does a process exist to require the review of quotes and orders to ensure that requirements are adequately defined?

- Is there a process for handling the review of verbal orders?

- Is there a process for handling the resolution of differences between quotations and orders?

- Is documented information retained of the results of reviews and actions taken?

Definitions (Refer to ISO 9000:2015)

- Documented information

- Requirement

- Review

Considerations for Documented Information to Be Maintained and/or Retained

- Clause 8.2 and its four subclauses have no specific requirement for maintaining documented information. Clause 7.5.1 requires the organization to determine the documented information to be maintained and retained. The organization should carefully consider how these reviews are accomplished, and we think this information should be maintained.

- Retention of documented information is required for the results of reviews of requirements.

Clause 8.2.4—Changes to Requirements for Products and Services

What Does the Clause Say?

Clause 8.2.4 states that changes are required to be controlled and documented information updated to ensure that changes are properly included in documented information.

What Does the Clause Mean?

When changes to product requirements, orders, contracts, or quotations occur, the organization is required to ensure that relevant documented information is amended and communicated, as appropriate, within the organization.

Consider the Potential Interactions as They Apply to Your QMS

• Clause 7.5 Documented information

• Clause 8.2.1 Customer communication

• Clause 8.2.2 Determination of requirements related to products and services

• Clause 8.3 Design and development of products and services

Implementation Tips

The simple and fundamental requirements indicated in "What Does the Clause Mean" are often much harder to meet in real life than to talk about on paper. Changes tend to come from all sorts of sources. Customer floor-level workers in today's environment often talk directly to factory workers in customers' plants. Cell phones are used to relate the latest changes to schedules and requirements. The situation can turn into chaos. Thus, control rules are needed so that decisions related to changes are made by the appropriate people with the relevant and up-to-date information. These considerations should be a key part of considering process interactions. Often rapid response is critical for the customer, so design the system in such a way that you can deliver just that!

Question to Ask to Assess Conformity

- Does a process exist for handling changes to product requirements?

Definitions (Refer to ISO 9000:2015)

- Documented information

- Product

- Requirement

- Service

Considerations for Documented Information to Be Maintained and/or Retained

Keeping good records of changes is both a challenge and a practical necessity.

Clause 8.3—Design and Development of Products and Services

What Does the Clause Say?

Clause 8.3, on design and development controls, has six subclauses:

- 8.3.1 General

- 8.3.2 Design and development planning

- 8.3.3 Design and development inputs

- 8.3.4 Design and development controls

- 8.3.5 Design and development outputs

- 8.3.6 Design and development changes

Design and development activities needed for products and services are required to be planned and controlled through an established, implemented, and maintained process. This process may be used for both products and services and for associated processes. It is required to include the following:

- *Planning* to determine design stages considering activities such as verification and validation, control of design interfaces, design review, resources needed for design and development, customer involvement, and the documented information needed to confirm that input requirements are met.

- Determination of the *design and development inputs* required, including such things as functional requirements, regulatory and statutory requirements, applicable standards or codes, information from earlier projects, and potential consequences of failure. Conflicting requirements are required to be resolved.

- *Design and development controls*, including clear delineation of the results to be achieved, planning and conducting design and development reviews and verification activities to ensure design outputs meet input requirements, and validation to ensure the products and services meet the requirement for the application intended.

- *Design and development outputs* are required to meet input requirements, to be adequate for subsequent processes in the provision of the product or service, and to ensure the products and services are fit for their intended purpose.

- *Design and development changes* are required to be identified, reviewed, and controlled. This includes changes to design inputs or outputs. Controls are required to ensure that changes do not have an impact on the products and service conformity.

The organization is required to *retain documented information* resulting from the design and development process, including design and development changes.

What Does the Clause Mean?

The requirements of clause 8.3 are essentially the same as in ISO 9001:2008, but the new version has been simplified a bit. The intent of clause 8.3.2 is to ensure that the organization plans and controls design and development projects. The key reason for this emphasis on planning is to maximize the probability that the project will meet defined requirements. If the design and development processes are well planned and controlled, an additional benefit should be that projects are completed on time and within budget.

Planning is required at the level of detail needed to achieve the design and development objectives—not to generate an excessive amount of paperwork. Stages of the project need to be determined, and responsibilities, authority, and interfaces need to be defined. Requirements need to be established for the incorporation of review, verification, and validation into the design and development project. The organization needs to determine how communications will be structured (e.g., weekly meetings, periodic reports, or other methods). In many cases a number of organizations are involved in this process, and the success of the design and development project often rests heavily on proper identification, understanding, and control of design interfaces.

One of the often quoted (and ludicrous) criticisms of ISO 9001 is that it can be used to ensure that a process is in place to produce conforming concrete life preservers. Whoever proffered this criticism did not understand the meaning of clause 8.3.3 (or its equivalent in ISO 9001:1994/2000/2008). This clause is intended to ensure the development and documentation of a requirements specification or an equivalent statement of the general and specific characteristics of a product to be developed, including the suitability of the product to meet marketplace and customer needs.

ISO 9001:2008 had a requirement that design and development output be provided in a way that can be used for subsequent verification. This provision has been deleted, but planning is required to take into account the needs of subsequent activities so the organization is not completely free to produce junk design outputs. This generally means there must be documented information that the design and development have been executed in accordance with the requirements defined at the inception of the project. This can be in the form of development reports with data to show the requirements have been satisfied. Such data or reports could include test results or any other formal documentation of the results of the effort to develop a product with the specified characteristics. In some cases the output may include mock-ups, models, or other means to communicate the intent of the design and development team.

Design and development review is required but is not covered in detail. This concept applies equally to hardware, processed materials, software, and service projects. In fact, it is a critical element of the software design and development process. When robust design and development reviews are held for software projects, including design and development reviews of software test plans, development cycles are typically reduced and life-cycle costs are lower.

Design and development review is intended to address the "abilities" associated with a new product—manufacturability, deliverability, testability, inspectability, shipability, serviceability, repairability, availability, and reliability, as well as issues related to inventory and production planning and the purchase of components and subassemblies. Design and development reviews are intended to identify issues, to discuss possible resolutions, and to determine appropriate follow-up.

The difference between design and development validation and verification has caused much confusion in the past, especially with new users of the standard. Both verification and validation are explicitly included in ISO 9001:2015.

Verification activities are conducted to ensure that the design and development outputs meet the input requirements.

Design and development validation is intended to ensure that the design and development output conforms to defined user needs and "is capable of meeting the requirements for the specified application or intended use, where known." Design and development validation is usually performed after successful design and development verification. It is worthwhile to state again the difference between verification and validation. *In simple language, verification addresses conformance to requirements, while validation addresses meeting defined user needs.*

In addition to documenting that the output results meet the input requirements, the standard requires that information be provided to facilitate product production and service delivery. For hardware products this means that the design and development team or individual should provide appropriate information to facilitate production of the product to specified requirements as well as providing valid input for design verification and validation. Providing clear product acceptance criteria from product design and development is essential for hardware, service, processed materials, and software products.

The output from the design and development process must specify the characteristics of the product that are essential to its safe and proper use. The output is expected to include any information that relates to producing or using the product safely and properly. Organizations should pay particular attention to this issue. It not only addresses ultimate customer satisfaction with the product or service, but also provides objective evidence that the organization considered the safe and proper use of products. The availability of such information could be important to demonstrating prudent judgment if there are liability issues related to the product or service. Conversely, not having such records could be viewed in litigation as evidence of a flawed design and development process.

Finally, the standard requires that the output from a design and development project be approved before the product is released. This requirement is included to ensure that all aspects of the project have been executed in accordance with documented plans and applicable procedures before the product is launched into production or delivered to a customer.

Consider the Potential Interactions as They Apply to Your QMS

- Clause 4.2 Understanding the needs of interested parties

- Clause 6.1 Actions to address risks and opportunities

- Clause 6.2 Quality objectives

- Clause 7.1.6 Organizational knowledge

- Clause 7.4 Communication

- Clause 8.1 Operational planning and control

- Clause 8.2.1 Customer communication

- Clause 8.2.3 Review of requirements for products and services

- Clauses 8.2.4, 8.3.6, 8.5.6, 9.3 Change to requirements customer expectations and organizational context

- Clause 8.4 Control of externally provided processes, products, and services

- Clause 8.5 Product and service provision

- Other processes that may not be considered part of the QMS, such as:

  — Innovation

  — Sales and marketing

  — Product pricing

  — Cost reduction

  — Capital allocation

Implementation Tips

A typical approach for *design and development planning* and control is to generate some form of project flowchart that incorporates the pertinent personnel, timing, and interrelationship information. Some simple examples include Gantt charts, program evaluation and review technique (PERT) charts, and critical path method (CPM) charts. With modern computing, new techniques abound. It is not grabbing for the latest buzzword that matters; rather, the important effort is the thinking and discussion required to determine how the project will proceed from inception to completion. Widely available software can be a useful tool in meeting the planning requirements of the standard (e.g., Microsoft Project, Primavera).

There are many areas to consider when defining product and service requirements and any other *design inputs*: statutory and regulatory requirements, environmental considerations such as ISO 14000, industry standards, national standards, International Standards, organizational standards, safety regulations, customer wants and needs, cost, past experiences, and, for designs that are related to specific customer orders, contract commitments.

The result of the consideration of such items is normally the documentation of a complete and unambiguous statement of product requirements, sometimes called a requirements specification. Development work should not begin until such a document exists in a form acceptable to all who have responsibility for contributing to the product specification (at least to those who must bring the product to the marketplace and those who must do the design and development). There is no requirement in ISO 9001:2015 for all parties to concur, but there may be contractual requirements to do so. Careful consideration of technical and commercial requirements can avoid misunderstandings during project implementation. It is especially worthwhile to obtain closure, where appropriate, between marketing or sales and those who will be doing the development work. A requirements specification signed by the involved parties is one way to ensure that concerned parties in an organization are in agreement regarding the product requirements. Such a document can provide the objective evidence that design inputs are understood. An important *design and development review* issue is to ensure the design community that design review will not interfere with the creativity and innovation of the designers or slow down the development process. Rather, it is a process step intended to provide confidence that the spectrum of internal and external customer needs has been considered as early as possible and addressed with the aim of ultimately ensuring external customer satisfaction while being cognizant of the time and budget restraints that exist for all development projects.

The standard does not prescribe the number of design and development reviews that should be conducted. This should be determined during the design and development planning process and should be modified, as appropriate, during the course of a project. Certainly one design and development review is a minimum unless conditions warrant formally waiving this requirement. In certain circumstances it may be appropriate to waive this requirement, in which case there should be documentation of the reasons and the authority for the waiver, which should be included in the project files. *Design review is a powerful tool to enhance the development of products and services that will be completed on time, within budget, and to internal and customer requirements.*

Verification is conducted to ensure the product is capable of meeting specified requirements and that objective evidence exists to demonstrate the basis for this assertion. This requirement applies equally to all product sectors. The service sector should be particularly attentive to conducting thoughtful product verification since there is usually not an opportunity to address nonconformity of product after it is delivered to a customer. Service, even more than hardware or software, must be right the first time to maximize the probability of customer satisfaction.

It is especially important to understand and address *validation* in the world of software product development because of the often mysterious interactions that occur deep in the workings of a computer. In addition to being an ISO 9001 requirement, and even though software designers complain there is never enough time to perform robust validation, this is not an area to be ignored or given perfunctory treatment.

For software, if the output from a project to design a unit or module of a software product (e.g., a statistical process control [SPC] package) performed as specified in the SPC requirements but caused a word processor to crash when the SPC product was loaded into a system, this product would "meet" the intent and requirement of the verification clause but not of the validation clause.

For hardware, if a water heater design meets all specified requirements but is not able to be easily installed by a plumber, it would "meet" the intent and requirement of the verification clause but not of the validation clause.

For service, the primary requirements might be met, but secondary factors can suddenly overshadow them. Validation helps uncover incomplete service requirements. For example, an express mail service that guarantees overnight delivery might meet the schedule, but it is inadequate if the package is left in a doorway during inclement weather and is destroyed by rain. This service would meet the intent and requirement of the verification clause but not of the validation clause.

In addition to the customer satisfaction implications, robust validation processes are critical to optimizing the life-cycle costs of software (the majority of which typically occur after product release) and minimizing product-liability exposure. Thus, validation should receive careful attention, and the results should be recorded and retained as records.

From the previous discussion, it is obvious that validation is performed after verification and in an environment that approximates as closely as possible the operating conditions that will exist in actual use. Also, whenever possible, it should be performed before the product is released for shipment. If it is not possible to perform complete validation of a hardware, software, or service product before release and/or shipment to customers, then validation should be performed to the extent that is reasonable and final validation performed when and as appropriate.

In addition to customer satisfaction issues, cost containment is a major reason for performing validation before a product is delivered to a customer. The resolution of issues after shipment can be very expensive.

The conventional documentation of the *outputs* of a design and development project demonstrates that the product will do what it is expected to do. A more difficult issue to address with regard to design and development output is how to document that the product will not do what it should not do. It is especially important, for example, to ensure that a software product will not interfere with the operation of other software. The documentation of the results of a development project is typically the responsibility of the team or person who performed the work on the project.

Design changes are often needed; after all, change is all around us! During the course of a design and development project, there are usually changes to the requirements that were defined at the input stage. Such changes occur for many reasons, including (1) omissions that become apparent after design and development work starts, (2) errors or inconsistencies in the design or in a specification requirement, (3) changes requested by marketing or by a customer, (4) perceived improvement opportunities, (5) changing regulatory or statutory conditions, (6) issues raised in design and development review, (7) issues raised during the verification process, and (8) issues raised during the validation process. Changes to the design and development project for any one or combination of reasons need to be addressed in accordance with the requirements of this clause. Much of the specificity that was in earlier versions of the design and development changes requirement has been deleted. But design and development changes must still be "controlled." It is prudent to not only review changes to ensure they do not compromise product or service conformity but also perform any additional verification and validation to ensure the change will meet intended outcomes. Failure to do this for a critical design change would certainly be a lack of the required "control."

A review of quality cost data will typically reveal the large costs that can be avoided by intense attention to control of all phases of design and development of products and services, in spite of organizational inertia to aggressively define and implement such controls.

Questions to Ask to Assess Conformity

8.3.1 General

- Is there a design and development process in place for situations where the organization is responsible for design and development?

- Is this process used and effective?

8.3.2 Design and Development Planning

- Are the stages of the design and development project defined? Where?

- Are appropriate design reviews addressed?

- Is the verification process addressed and is it appropriate?

- Is validation addressed for all applications as appropriate?

- Is it clear who is responsible for what?

- Are the communications channels and interfaces defined and managed?

- Is there evidence that communication on projects is occurring and that it is effective?

- Does the planning define the documented information to be retained? To be maintained?

8.3.3 Design and Development Inputs

- Are requirements for new products defined and is documented information (i.e., records) maintained?

- Are the requirements complete?

- Are the requirements unambiguous?

- Are the requirements without conflict?

- Is there an effective process to resolve design issues, conflicts, and interface problems?

8.3.4 Design and Development Controls

- Does the design and development planning include careful consideration of the design and development controls needed—the timing, personnel, equipment, and other resources necessary to carry out the controls?

- Are design and development reviews being performed?

- Are design and development reviews indicated in the project planning documents?

- Who attends design reviews?

- Is the attendance appropriate?

- Are results documented?

- Are follow-up actions taken?

- Are appropriate records maintained?

- Is a verification process in place?

- Is the verification process effectively implemented?

- Is design and development validation performed to confirm that the product is capable of meeting the requirements for intended use?

- Are suitable controls provided in cases where full validation cannot be performed prior to delivery?

- Is documented information (i.e., records) of design and development validation maintained?

8.3.5 Design and Development Outputs

- Is the output of design and development projects in a form suitable for verification against inputs?

- Does the design and development output satisfy input requirements (e.g., as stated in a functional requirements specification)?

- Does output provide, as appropriate, information for purchasing, production operations, and service provision?

- Are product acceptance criteria clearly stated?

- Are product safety and use characteristics identified?

- Is there an approval process for the release of products from the design and development process?

- Are all design and development project changes reviewed to ensure there is no adverse impact on conformity to requirements?

8.3.6 Design and Development Changes

- Are all design and development project changes documented and reviewed?

- Are design and development changes verified and validated, as appropriate?

- Is there evidence to demonstrate that changes are authorized?

- Do records include the results of reviews of changes?

- Have changes been communicated to interested parties?

- Do records include follow-up actions related to the review of changes?

Definitions (Refer to ISO 9000:2015)

- Design and development

- Monitoring

- Objective

- Output

- Regulatory requirement

- Requirement

- Review

- Specification

- Statutory requirement

Considerations for Documented Information to Be Maintained and/or Retained

- Clauses 8.3.3, 8.3.4, and 8.3.5 require the organization to maintain documented information on the design inputs, controls, and outputs to confirm that design and development requirements have been met

- Clause 8.3.6 has detailed requirements for documented information retention related to changes

- We recommend that the organization also maintain documented information (both procedures and records) on all elements of the design and development process, as appropriate

Clause 8.4—Control of Externally Provided Processes, Products, and Services

What Does the Clause Say?

Clause 8.4 has three subclauses:

- 8.4.1 General

- 8.4.2 Type and extent of control

- 8.4.3 Information for external providers

Control requirements for externally provided products and services are required to be determined by the organization when:

- The externally provided product or service becomes part of the organization's product or service

- External providers provide product or service directly to the customer on behalf of the organization

- A process or part thereof is outsourced by the organization

In these circumstances, the organization is required to use criteria for the evaluation, selection, monitoring of performance, and reevaluation of external providers. This is to be based on the provider's ability to provide conforming products and services.

The organization is required to retain documented information on the results of the required evaluation, and any actions required as a result of the evaluations. The organization is also required to determine the controls to be applied based on the potential impact of the externally provided items on the organization's ability to consistently meet requirements. Verification or other activities are required to be applied as necessary. The effectiveness of the controls applied by the outside provider must also be considered.

Applicable requirements shall be communicated to external providers for:

- The processes, products, and services to be provided
- Release approval for products and services and any associated process releases
- Any required methods, processes, and equipment
- Competence and qualification of personnel
- Interactions with the organization's quality and management systems controls and monitoring to be applied to the external provider
- Intended verification or validation activities at the external provider's location by the organization or its customer

The organization is required to ensure the adequacy of this information before providing it to the outside provider.

What Does the Clause Mean?

The requirements of this clause are similar to those of ISO 9001:2008 clause 4.1 related to outsourcing and clause 7.4 on purchasing. The two concepts—control of purchasing and control of outsourced products, services, and processes—are combined, so the wording has been generalized to cover all products, services, and processes of external providers.

Clause 8.4 requires controls for products or services of external origin, whether you are dealing with a supplier, another entity of your own organization, a subcontractor, a partner, or an outsourcing of parts, services, or processes. It does not matter what you call it, the control requirements are the same. When "outsourcing" of processes was first addressed in ISO 9001:2000, a great deal of angst occurred because organizations were often not using normal purchasing control for outsourced work. ISO 9001:2015 makes it clear that while different people in the organization may have different processes for buying things, they must all conform to the same basic requirements if:

- The externally provided product or service becomes part of the organization's product or service

- External providers provide product or service directly to the customer on behalf of the organization

- An entire process or function is outsourced by the organization (e.g., design, after-delivery service)

A single organization may utilize differing controls for outsourcing processes, buying raw materials, or buying items to drop-ship to customers. The controls may be different, they may be executed by different parts of the organization, but they must all conform to ISO 9001:2015 clause 8.4.

The actual controls to be applied and the extent of those controls are to be determined by the organization for each situation. In making the determination, the organization is required to take into account:

- The potential impact of the externally provided product or service on its ability to provide customers with conforming product, and the extent of controls

- The perceived effectiveness of the external provider's controls

The requirement for verification of externally provided processes, products, and services has been greatly simplified. But the intent remains essentially the same as that of ISO 9001:2008. The organization also needs to determine:

- The type and extent of verification to be applied

- Where, when, and by whom that verification is to be conducted

The external provider must receive any appropriate information it requires. Clause 8.4.3 has a list of items that need to be included. That list may not be complete, so the clause requires the organization to make certain the information is adequate prior to communicating it to the external provider.

A key message in the new standard is that products or services of external origin are within the scope of the organization's QMS regardless of who provides it or the means of delivery.

Consider the Potential Interactions as They Apply to Your QMS

- Clause 4 Context of the organization

- Clause 5 Leadership

- Clauses 6 and 8 Quality management system and its processes

- Clause 8.3 Design and development of products and services

- Clause 8 Operational planning and control

- Clauses 6.1, 9, 10.2 Actions to address risks and opportunities

- Clause 8 Operational roles, responsibilities, and authorities

- Clause 8.7 Control of nonconforming process outputs, products, and services

- Clause 5 Finance processes

- Clauses 6, 8, 9.2, 9.3 Cost reduction processes

- Clauses 6, 8, 10 Programs for supply base reduction or multiple sourcing

Implementation Tips

The organization needs to make certain it has appropriate controls to cover all items from external sources. Innovative new ways of external provision are being invented regularly, and the QMS needs to ensure controls are adequate. Organizations need to develop processes that ensure adequate evaluation, selection, and control of externally provided products and services. Clause 8.4 of the new standard does not talk about these processes, but remember, clause 4.4 on the QMS and its processes is applicable to this subject, as is clause 8.1.

It is often useful to work from a list of potential external providers developed with the input of staff from appropriate functions of the organization. This list may be filtered and reduced to a few likely candidates using tools such as request for quotation, survey, submission and analysis of samples, and quality systems audit. Selection of external providers should be based on a review of their abilities to meet business and technical requirements for quality, cost, delivery, and other considerations important to the organization. Competitive comparisons of financial, business, and quality attributes should be considered where applicable.

At a minimum, information communicated to external providers needs to include the items listed in clause 8.4.3, *Information for external providers*. But there may be a number of other things to consider, such as:

- Does your organization need confidentiality and nondisclosure agreements not otherwise contained in contracts?

- Do you need a statement of work?

- Based on criticality of the product, service, or process, do you need to develop a quality agreement with the external provider?

- What expectations do you have for your external provider with regard to risk-based thinking? Keep it simple but consider what you need to agree on in writing with your external provider to ensure effective and efficient conformity to requirements.

- To what extent do you need agreement with your external providers on their internal product or service production and delivery processes?

- Do you need agreement on requirements to control the external provider's product, service, or process changes?

- Have you made clear your expectations of external providers to perform corrective actions?

- Have you considered product liability exposure and addressed the implications for documented information to demonstrate that your organization has exercised prudent judgment to ensure conformity of products and services to requirements?

Use careful judgment when deciding what verification you will perform of externally provided products, services, and processes. Many organizations focus on developing supplier relationships that treat suppliers as an extension of the organization. This enables the organizations to forgo much of the old thinking of receiving inspection, source visits, and so on. In the "old days," verifying that purchased product met specified purchase requirements was done through receiving inspection. The higher the quality, the lower the inspection needed in order to have confidence in the quality of the supplied product. For some purchased products, this approach remains viable. On the other hand, in today's complex world, the process approach emphasizes that success with external providers and obtaining flawless processes, products, and services normally require much more than this historical tollgate approach. Often this involves suppliers' demonstration of process capability and control.

Think of the total system including the entire supply chain. The performance of external providers is more critical today than ever before. The performance of externally provided processes, product, and services needs to be included in your organization's overall system for managing quality, along with:

- Identification of and dealing with risks related to external providers' performance

- Quality objectives

- Audit results

- Measurement and analysis of data

- Corrective actions, improvement, and management review

Consider also what you need to do to foster stronger relationships with external providers. Often, communication to external providers of their performance and the results of your monitoring are key interface elements.

Questions to Ask to Assess Conformity

- Has the organization developed a process or processes for control of externally provided products, services, and processes?

- Are these processes comprehensive and effective?

- Do the controls apply to products and services provided by external providers for incorporation into the organization's own products and services?

- Do the controls apply to products and services provided directly to the organization's customers by external provider?

- Do the controls apply to a process or part of a process that is provided by an external provider?

- Are the products, processes, and services adequately described to the external provider?

- Is the external provider given adequate information on approval or release of products and services, methods, processes, or equipment as applicable?

- Is the external provider given adequate information on requirements for personnel competence and qualifications as needed for the application?

- Is the external provider given adequate information for interacting with the organization's QMS and other management systems?

- Is the external provider given adequate information on the control and monitoring to be applied?

- Is the external provider given adequate information on verification activities the organization or its customers may perform?

- Has the organization defined a process for verifying that purchased product conforms to defined requirements?

- Is the process effectively implemented?

- Is documented information retained as required?

- Have the appropriate linkages with other processes of the QMS been determined, and are they managed?

Definitions (Refer to ISO 9000:2015)

- Audit

- Change control

- Design and development

- Determination

- Deviation permit

- Document

- Documented information

- Information

- Interested party

- Knowledge

- Management system

- Monitoring

- Objective

- Organization

- Output

- Outsource

- Product

- Provider

- Quality characteristic

- Quality management system

- Quality planning

- Regulatory requirement

- Requirement

- Review

- Risk

- Service

- Specification

- Statutory requirement

- Supplier

- System

- Validation

- Verification

Considerations for Documented Information to Be Maintained and/or Retained

- Clause 8.4.1 requires the organization to retain documented information on selection, monitoring, evaluation, and reevaluation of external providers

- We recommend that the organization also retain documented information on the data and information it provides to external providers

- We also recommend that the organization maintain documented information on the processes for control of externally provided products, services, and processes

PART 2: CLAUSES 8.5, 8.6, AND 8.7— WHERE THE RUBBER MEETS THE ROAD

Part 1 covered clauses 8.1–8.4. We will now address the last three subclauses of clause 8, which are:

- 8.5 Production and service provision
  - — 8.5.1 Control of production and service provision
  - — 8.5.2 Identification and traceability
  - — 8.5.3 Property belonging to customers or external providers
  - — 8.5.4 Preservation
  - — 8.5.5 Post-delivery activities
  - — 8.5.6 Control of changes
- 8.6 Release of products and services
- 8.7 Control of nonconforming outputs

These clauses involve actual production of products or delivery of services. They include such activities as making, delivering, and supporting products and services after delivery, control of nonconformities, and release of products and services for delivery to the customer. See the bottom portion of Figure 7.2.

Clause 8.5—Production and Service Provision

What Does the Clause Say?

The ultimate requirement for this part of ISO 9001:2015 can be thought of as applying to the whole standard: Conduct all

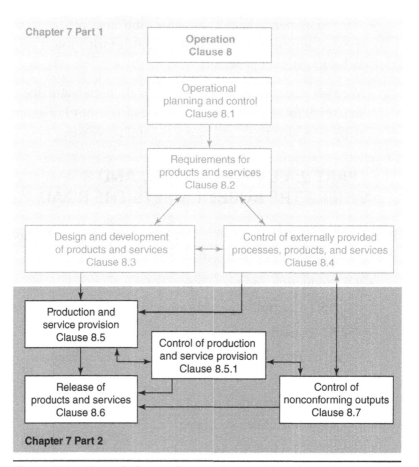

Figure 7.2 Controls for product service provision through release of products and services.

work and perform all processes under controlled conditions. For product and service provision, the state of control required includes:

- Documented information to define the characteristics of the products and services, the activities to be performed, and the results to be achieved

- Monitoring, measurement, and process control activities to ensure that outputs meet acceptance criteria

- Controlled use of needed infrastructure and process environment

- Competent personnel with the required qualifications and implementation of any actions required to prevent human error

- Process validation and periodic revalidation to ensure output requirements are met for processes with outputs that cannot be verified

- Implementation of actions to prevent human error

- Implementation of product and service release and post-release activities

Suitable identification is required where necessary. Where there is a traceability requirement, the organization is required to provide unique identification and to retain documented information to maintain the required traceability. Monitoring and measurement status is required to be maintained for outputs.

The organization is required to exercise care for property under its control that belongs to its customers or to its external providers. This property is required to be identified, verified, protected, and safeguarded. If such property is lost, damaged, or found unsuitable, the organization is required to report this to the owner and retain documented information on the occurrence.

The organization is required to preserve process outputs during production and service provision to meet applicable requirements.

Any requirements for post-delivery activities must be met. The organization is also required to consider the following issues:

- Customer requirements and feedback

- Nature, use, and intended lifetime of products and services as well as any potential undesired consequences

- Statutory and regulatory requirements applying to the product or service

The organization is required to review and control unplanned change essential for production or service provision to ensure ongoing conformity. Documented information is required on the reviews and actions taken.

Planned arrangements are required for release of the products and services to the customer. Release cannot take place until required product verifications are complete and issues related to product conformity have been resolved.

What Does the Clause Mean?

Clause 8.5 of the new standard is very nearly identical to the requirements of ISO 9001:2008 clause 7.5 of the same title.

The requirements are expressed in general terms, and their meaning should be clear for software and service providers. Although these requirements use very general language, they should not be interpreted as lessened or weakened requirements for hardware producers or providers of processed materials.

The focus of clause 8.5.1, *Control of production and service provision, is a key concept that processes need to be carried out under controlled conditions.* Considerations for achieving controlled conditions are presented. The requirements to determine the extent to which production and service operations are planned, established, documented, verified, and validated are presented in clause 8.1. Since clause 8.1, *Operational planning and control*, applies to all of the operations processes, these planning and development requirements are not repeated. *Controls for production and service delivery must be planned!*

When the processes are such that achievement of the product specifications cannot be fully verified by examining the finished product, either at an earlier stage of production or at the end, process validation must be performed. The inability to fully verify every unit of product may be due to the nature of the testing (e.g., when the testing is destructive). In such cases, the process must be validated. These requirements were spelled out in detail in ISO 9001:2008 clause 7.5.2, *Verification of processes for production and service provision.* These requirements are greatly simplified in the new standard and are incorporated into clause 8.5.1f.

After being validated, processes must be maintained in a validated state. The requirement in ISO 9001:2008 clause 7.5.2, *Defining the conditions and criteria for revalidation,* has been deleted, but the requirement for revalidation has been retained. Recall that planning and control of all operations was already covered in 8.1, *Operational planning and control.* If changes are made to the process equipment, the product design, the materials used to produce the product, or to other significant factors such as new personnel, the process often requires revalidation. The organization therefore needs to define the conditions that require a revalidation to be performed. Even if no initiating events occur, common practice often requires revalidation after a minimum period of use, a number of operating cycles, or a period of time. In some industries, revalidation is required if a period of five years has elapsed without an intervening validation. Validation should be carried out at appropriate intervals to respond to changes in market requirements, regulations, or standards in addition to ensuring the continued acceptable performance of processes.

For services, examples of processes requiring validation include those that create financial or legal documents and those that deal with professional advice. Validation includes considering a number of factors such as the need to qualify the processing method, the qualifying of service equipment, and having competent and qualified personnel providing the service.

For hardware and processed material, examples of processes that require validation include welding, soldering, gluing, casting, forging, heat-treating, and forming. Products with quality characteristics that require certain test and inspection techniques for verification, such as nondestructive examinations (e.g., radiographic, eddy current, or ultrasonic examination), environmental testing, or mechanical stress tests, usually require process validation. Validation includes consideration of the qualification of equipment, personnel, and processes.

For software, the organization may not be able to verify even the simplest code through testing. The expectations are that all software needs to be created by controlled processes following the model described in clause 8.3. The methods and extent of the process validations will differ widely based on the criticality and use of the software. The qualification of personnel, equipment, and software development methodologies and procedures is an important aspect of ensuring that software conforms to specified requirements. Many organizations go well beyond the requirements of ISO 9001 to control software processes due to the cost and customer satisfaction issues related to the release of software that does not meet customer requirements or expectations.

The requirements of ISO 9001:2015 clause 8.5.2, *Identification and traceability*, are the same as ISO 9001:2008 clause 7.5.3 of the same title. Minor wording changes have been made to clarify intent. The note in the 2008 version related to configuration management has been deleted, as this only applied in some more complex industries. We strongly recommend that configuration management not be ignored when planning and implementing your QMS.

Identification and traceability are separate but related issues. The degree of product identification that is needed must be determined, including any requirements for tracking purchased components, materials, and supplies that are uniquely related to the product. Appropriateness depends on the nature of the product, the nature and complexity of the process, industry practice, and whether identification is required in a contract. Integral to product identification is its status in meeting requirements at various stages of production or service development, storage, and delivery as indicated by passing tests and inspections.

In order to trace a product, the organization must identify the product and its component parts in adequate detail. Thus, traceability is closely related to identification. Full traceability involves the ability to trace the history, application, and location of an item or activity. It is used when there is a need to trace a problem back to its source and when it is necessary to be able to isolate all parts of an affected batch.

Documented information needed to ensure traceability should be defined. For example, traceability is typically a contract requirement for certain medical devices, defense/space vehicle assemblies, and many other applications.

The requirements of clause 8.5.3, *Property belonging to customers or external providers*, are essentially the same as those of ISO 9001:2008 clause 7.5.4, *Customer property*, except the clause now applies to property of external providers as well as customers. Very minor changes were made for clarity. This property is owned by the customer or an external provider and furnished to the organization for use in meeting the requirements of the agreement between the two. Upon receipt of the property from the customer, the organization agrees to safeguard the property while it is in the organization's possession. The requirements for the control include all property provided by the customer or external provider, including such items as tooling, information, test software, and shipping containers. It is fundamental that the customer provide product that is acceptable for the intended purpose. However, if this is product that needs to be repaired, for example, it would need to be acceptable for repair. If not, the organization might return the product to the customer as not repairable. A contractual business relationship, written or understood, should deal with this situation. The note in clause 8.5.3 makes it clear that property belonging to customers and external providers is considered broadly. For example, information or other intellectual property is a type of property subject to these controls.

The requirements for preservation are simplified from those of ISO 9001:2008 clause 7.5.5, but the intent remains the same. The organization must safeguard and protect the product during and between all processing steps through to delivery. It should have a system for appropriately identifying, handling, packaging, storing, and delivering the product, including its components. More attention is provided to post-delivery activities in ISO 9001:2015 than in ISO 9001:2008, but the requirements are self-evident.

A clause new to the area of product production and service delivery requires control of planned changes and review of the consequences of unintended changes (in clause 8.1). While the specific requirement may be viewed as new, the same concept applies to the whole QMS and is covered in ISO 9001:2008, under QMS planning in clause 5.4.2, which required the integrity of the QMS be maintained during change, and in ISO 9001: 2015 clause 6.3, *Planning of change*, requiring that all QMS changes be planned.

Consider the Potential Interactions as They Apply to Your QMS

- Clause 4 Context of the organization

- Clause 8.1 Operational planning and control

- Clause 8.2 Requirements related to products and services

- Clause 8.3 Design and development of products and services

- Clause 8.4 Control of externally provided processes, products, and services

- Clause 8.6 Release of products and services

- Clause 8.7 Control of nonconforming outputs

- Resources, particularly:

— Clause 7.1.2 People

— Clause 7.1.3 Infrastructure

— Clause 7.1.5 Monitoring and measuring resources

— Clause 7.1.6 Organizational knowledge

— Clause 7.2 Competence

- Clause 9 Performance evaluation

- Clause 10 Improvement

Implementation Tips

8.5.1—Control of Production and Service Provision

The organization should control operations processes by considering a number of factors. The organization must determine the production and service processes that need to be controlled and the outputs that must be achieved at each stage of processing. Also, the specific items of equipment that are needed to achieve the product specifications, including their sequence and operating conditions, need to be addressed.

The organization needs to determine the criteria of acceptability for these processes and perform evaluations against these criteria. One approach is to perform capability studies to demonstrate process suitability. From the determination of suitability, decisions can be made whether additional process development effort is required to achieve the product specifications with consistency. Appropriate criteria and controls for these processes need to be determined and implemented to maintain process capability and to prevent nonconformities.

The corresponding documented information (e.g., procedures, work instructions) needs to be maintained, and the associated measurement equipment should be identified. In determining the extent of documented information needed, the organization should consider the competency of its people, its size and complexity, and the criticality of the product. Controls often include direct measurement of process parameters and characteristics. There may be inspection or test of the output of the process (e.g., the product or service). In some cases, both are used. Whichever combination of approaches is adopted, verification activities should be integrated to maximize both the efficiency of the verification and confidence in the product. This clause requires organizations to think about their operations.

When assessing the conformity of processes to ISO 9001: 2015 requirements, it is helpful for organizations to keep in mind that *virtually all other requirements of the standard can be ignored if one can attest an affirmative belief that a process being appraised for conformity to requirements was planned, is being carried out, and is considering improvement under controlled conditions.* It is also helpful to keep in mind that the most astute external auditors also look for similar evidence.

For services, every organization should be aware of the requirements and conditions for the proper operation of its planned and offered services and should establish these in writing. One approach is to rank the offered services in order of their importance, cost, and criticality. The task of the organization is to plan, monitor, and systematically supervise the fulfillment of services so that the quality objectives can be achieved. For some types of services, there is little or no process equipment to control, as the service consists of actions performed by the service personnel directly for or with the customer. In these cases, the requirements of this clause apply directly to the service personnel competence and to the processes that establish the service delivery activities to be performed (see Chapter 6). Although it may be difficult to establish criteria for acceptable service delivery, some criteria are desirable so conformity of service delivery can be judged on a basis other than one that is purely subjective and possibly inconsistent, especially if services are being delivered by many individuals.

For hardware and processed materials, raw materials, parts, and subassemblies need to conform to appropriate specifications before being introduced into processing. In determining the amount of testing or inspection necessary, organizations should compare the cost of evaluation at various stages with the added value of subsequent activities. The economic impact of discarding or reworking product should be considered when planning the controls. In-process materials should be appropriately stored, segregated, handled, and protected to maintain their suitability. Special consideration should be given to shelf life and the potential for deterioration. Where in-plant traceability of material is important to quality, appropriate identification should be maintained throughout processing to ensure traceability to original material identification and quality status. Where important to quality characteristics, auxiliary materials and utilities such as water, compressed air, electric power, and chemicals used for processing should be controlled and verified periodically to ensure uniformity of effect on the process. Where an element of the work environment (such as temperature, humidity, or cleanliness) is important to product and service quality, appropriate limits should be specified, controlled, and verified (see clause 7.1.4 in Chapter 6).

For software, the creation of code is usually part of the design and development process. Mass production of software involves replicating the code and may involve consulting services to tailor the software to specific customer needs. Nevertheless, it is essential to control the final stages of development through to code replication and subsequent installation and servicing processes as well as the activities related to preparing and delivering the "media" (i.e., the software code) to users.

In the area of identification and traceability, configuration management is very important for software, aviation, medical devices, and other situations. The process controls for any required configuration management activities should include maintenance of documented information for configuration management to the extent appropriate. This discipline should have been initiated early in the design phase and should continue throughout the life cycle of the software. It assists in the control of design, development, provision, and use of the software, and it enables the organization to examine the state of the software during its life. Configuration management can include configuration identification, configuration control, configuration status accounting, and configuration audit.

It is not uncommon for organizations to provide a combination of hardware, processed materials, services, and software to customers. The controls for each product category may be different, but their provision to a customer must be achieved in an integrated way.

8.5.2—Identification and Traceability

It may be important to identify specific personnel involved in each phase of a service delivery process. Different personnel may be involved in successive service functions, each of which is to be traceable. For example, the recording through signatures on serially numbered documents in banking operations is often required. In this case, there is no tangible product, but each person's identity needs to be traceable to provide the appropriate documentation trail. In a different application, signatures often serve as indications of processing status and approval to proceed with payment of, for example, an invoice.

For hardware and processed materials, product identification is often achieved by marking or tagging a product or its container. When products are visually identical but their functional characteristics are different, different markings or colors may be used. More often, quantities of product are segregated into batches with unique batch numbers. Batch definition may be determined by identifying potential sources of batch-to-batch variation. Sources of variation traditionally are the five Ms: man (operator), method (or procedure), material, measurement (measurement method and instrument), and machine (or processing step). New batches may be defined as these potential sources change. Traceability approaches vary widely and depend on the application regarding how or if the unique identification of product is accomplished.

For software, configuration management practices require that each version of a configuration item be identified by some appropriate means. Likewise, there is a need to maintain the status of the verification steps and tests that have been completed. The results achieved by the product or product components at each phase of the development cycle must also be maintained.

8.5.3—Property Belonging to Customers or External Providers

For services, this requirement entails a service provided by the organization to the customer. Repair of a piece of equipment requires clearly defining the responsibilities of both parties. An automotive repair shop must not damage a customer's car. On a much larger scale, shipowners contract for the repair of ships with private shipbuilders using owner-furnished equipment and owner-furnished material for a ship that often is to arrive at some future date. Equipment and material may be held in inventory before and after repairs are completed by the private shipbuilder. The storage and handling of supplied material must be considered.

For hardware and processed materials, the organization should examine property received from a customer for identity, quantity, and damage. The property should be safeguarded and maintained. The organization may need to provide maintenance or utilize a maintenance contract with a third party. In such cases, the contractual agreements need to be clear as to responsibility.

For software, this clause can be a significant factor. An example is the case where a customer provides source code to a contract programming organization for modification to incorporate additional features. The organization must exercise care in protecting the original functionality of the software and unauthorized access to it. Detailed agreements typically define these relationships, including verification and validation requirements of the changes and the protection of intellectual property of the customer.

8.5.4—Preservation

Marking and labeling should be readable, visually or by machine. Consideration should be given to processes for segregating batches, stock rotation, and expiration dates. Packaging, containers, wraps, and pallets should be appropriate and durable for protecting the product from damage. Suitable storage facilities that include both physical security and protection from the environment should be provided. It may be necessary to check product periodically for deterioration. The organization should provide appropriate handling and transportation equipment, such as conveyors, vessels, tanks, pipelines, or vehicles, to minimize harm due to handling or to exposure to the environment.

For services, this clause can be viewed as applying to the physical things involved in the service delivery. Examples include delivery of packages by air freight, trucking services, and food-delivery services. It is the carrier's responsibility to preserve the condition of the goods it carries, and the organization's responsibility to ensure that any transportation services used has the capability to protect the products delivered to customers.

8.5.5—Post-delivery Activities

For activities like warranty work and post-delivery services on equipment, the organization should plan the applicable activities and controls to ensure conformity to requirements.

8.5.6—Control of Changes

Changes, regardless of their source or cause, need to be controlled. This means establishing a state of control. See the list of items needed to establish such controlled conditions in clause 8.5.1. Sometimes this means establishing a state of control in the face of a chaotic product or service delivery situation. In the midst of chaos, it is often easy to "forget" some of the key control elements. A strong, disciplined approach will generally turn out best in the long run. A defined process that provides change control requirements is recommended.

Questions to Ask to Assess Conformity

8.5.1 Product and Service Provision

• Are specifications available that define quality characteristic requirements of the product or service?

• Has the organization determined the criteria of acceptability for demonstrating the suitability of equipment for production and service operations to meet product or service specifications?

• Has the organization demonstrated the suitability of equipment for production and service operations to meet product or service specifications?

• Has the organization defined all production and service provision activities that require control, including those that need ongoing monitoring, work instructions, or special controls?

• Are work instructions available and adequate to permit control of the appropriate operations to ensure conformity of the product or service?

• Have the requirements for the work environment needed to ensure the conformity of the product or service been defined, and are these work environment requirements being met?

- Is suitable monitoring and measurement equipment available when and where necessary to ensure conformity of the product or service?

- Have monitoring and measurement activities been planned, and are they carried out as required?

- For hardware, processed material, and software, have suitable processes been implemented for release of the product and for its delivery to the customer?

- Have suitable release mechanisms been put in place to ensure that product and service conform to requirements?

- Has the organization determined which production or service processes require validation? Have these processes been validated?

- Does the organization use defined methods and procedures to validate processes?

- Have the requirements for retained documented information of validated processes been defined?

- Is the required documented information of validated processes maintained?

- Have the processes requiring revalidation been defined?

- Have processes, as required, been revalidated?

- Is adequate documented information maintained to ensure that process validation is effectively implemented?

8.5.2 Identification and Traceability

- Has the product been identified by suitable means throughout production and service operations?

- Has the status of the product been identified at suitable stages with respect to monitoring and measurement requirements?

- Is traceability a requirement?

- Where traceability is a requirement, is the unique identification of the product recorded and controlled?

8.5.3 Property Belonging to Customers and External Providers

- Has the organization identified, verified, protected, and maintained property owned by the customer or external provider that is provided for incorporation into the product?

- Does control extend to all customer property, including intellectual property?

- Can the organization demonstrate that the customer or external provider has been or will be notified if the property has been lost, damaged, or otherwise found to be unsuitable?

8.5.4 Preservation

- Does the organization handle the product during internal processing and delivery so as to preserve conformity to customer requirements?

- Does the organization store the product during internal processing and delivery so as to preserve conformity to requirements?

- Does the organization protect the product during internal processing and delivery so as to preserve conformity to requirements?

8.5.5 Post-delivery Activities

- Are the risks associated with products and services considered?

- Are the nature and intended lifetime of the product or service considered when planning post-delivery activities?

- Are statutory and regulatory requirements considered?

- Is customer feedback considered?

- Is there evidence that post-delivery activities are planned to meet requirements?

- Are post-delivery requirements met?

8.5.6 Control of Changes

- Are unplanned changes essential to meeting requirements reviewed?

- Are changes and their effects controlled?

- Is documented information retained on review results, the person authorizing the change, and any necessary actions?

Definitions (Refer to ISO 9000:2015)

- Capability

- Change control

- Characteristic

- Customer

- External provider

- Measurement process

- Measuring equipment

- Process

- Quality characteristic

- Release

- Requirement

- Risk

- Specification
- Traceability
- Validation
- Verification
- Work environment

Clause 8.6—Release of Products and Services

What Does the Clause Say?

Planned arrangements are required at appropriate stages to verify that products and services conform to requirements. The organization is required to retain documented information of the conformity with acceptance criteria.

Products and services shall not be released to the customer until these planned arrangements have been completed or unless otherwise approved by relevant authority and if required by the customer. Documented information is required to be retained as evidence of conformity and traceability to personnel authorizing release.

What Does the Clause Mean?

This clause contains the requirements for product release found in ISO 9001:2008 clause 8.2.4, *Monitoring and measurement of product*. The requirements have not changed. Stage-by-stage determination of conformity is permitted. The final determination needs to confirm that the product conforms to acceptance criteria.

Consider the Potential Interactions as They Apply to Your QMS

- Clauses 6 and 8 Operational planning and control
- Clause 7 Support processes
- Clause 8 Determination of requirements for products and services

- Clause 8.5 Product and service provision
- Clause 8.3 Design and development
- Clauses 8.7 and 10.2 Control of nonconforming outputs
- Clause 9 Performance evaluation
- Clause 10 Improvement

Implementation Tips

Conformity to requirements needs to be verified, and documented information needs to be retained to substantiate conformity and to indicate who within the organization authorized release of final products and services.

Product release and service delivery require that all specified activities be accomplished unless release is otherwise approved by a relevant authority or by the customer. This usually means that some form of documented information (i.e., a record) should be available to document that specified activities have been accomplished.

As appropriate, release methods need to be developed and implemented before providing the service to the customer. Release methods differ in form, timing, and application. For example, airline pilots use preflight checklists to verify that requirements have been met prior to takeoff. An automobile repair shop uses both test instruments and a test drive to verify the satisfactory completion of its service before releasing a repaired vehicle to the customer. Also, customer-contact employees can receive immediate feedback by asking customers if the services have been adequately provided.

Questions to Ask to Assess Conformity

- Is there objective evidence that acceptance criteria for the product have been met?
- Do records identify the person authorizing release of the product?

- Are all specified activities performed before product release and service delivery?

- If there are instances in which all specified activities have not been performed before product release or service delivery, has a relevant authority (or as appropriate, the customer) been informed and has this individual approved the action?

Definitions (Refer to ISO 9000:2015)

- Conformity

- Nonconformity

- Objective evidence

- Product

- Release

- Service

- Verification

Clause 8.7—Control of Nonconforming Outputs

What Does the Clause Say?

Identification and control are required for process outputs, products, and services that do not conform to requirements. This control is to prevent their unintended use or delivery. The organization is required to deal with the nonconformity by one or more of the following as applicable:

- Correction of the nonconformity

- Segregation, containment, return, or suspension of provision of products and services

- Obtaining authority to use "as is," release, continuation, or re-provision of products and services, and acceptance under concession

If the nonconformity is subjected to correction, the organization is required to verify final conformity. The organization shall take appropriate corrective action on nonconformities in process outputs, products, and services, including those detected after delivery.

Retention of documented information is required of actions taken and any concessions obtained. This documented information is required to identify the person or authority who decided how to deal with the nonconformity.

What Does the Clause Mean?

This clause is structured and worded much differently than clause 8.3, *Control of nonconforming product*, of ISO 9001:2008. The structure and wording have been simplified. On the other hand, the intent and basic requirements have been the same since the 1987 version. From the beginning, the intent has been to prevent inadvertent delivery, use, or installation of nonconforming product. The new wording specifically addresses control of nonconforming process outputs and services. While this has been the intent all along, it is now clearly stated. Many "old-timers" will remember material review boards that were frequently convened to handle the disposition of nonconforming material. As our organizations have migrated to more service-oriented products, the relevance of this clause has diminished somewhat, although the intent is still applicable to hardware. For example, how does an organization control the delivery of a nonconforming service? Often this is not possible. While ISO 9001:2008 required a documented procedure for this control, the new version requires only retention of documented information regarding the decisions made and who made them.

This clause states actions to be considered when dealing with nonconformity. We recommend that when addressing nonconformity and clause 8.7 requirements that clause 10.2 be considered for interaction implications.

Consider the Potential Interactions as They Apply to Your QMS

- Clause 7.2 Competency

- Clause 8.1 Operational planning and control

- Clause 8.3 Design and development

- Clause 8.5 Product and service provision

- Clause 8.6 Release of products and services

- Clauses 8.7 and 10.2 Nonconformity and corrective action

- Clause 10 Improvement

Implementation Tips

A primary concept behind clause 8.7 is to ensure the effective implementation of processes that prevent or control unintended (or even intended) use or delivery of process outputs, products, or services that do not conform to requirements. This is a simple requirement that makes business sense. The challenge to an organization is to devise processes to accomplish this objective in a way that encourages personnel to address nonconformity of product rather than to find ways to avoid identification and control of nonconformities.

Another foundational concept used in an earlier version of ISO 9001 is that the organization needs to deal directly with nonconformities that are discovered after delivery of the product or service. ISO 9001:2015 requires that correction be extended to this situation, but not corrective action. We consider this to be an error in thinking as consideration of product recall is a significant concern. Clause 10.2.1 requires consideration of both correction and corrective action for nonconforming conditions, and we believe that such a process is preferred.

Typically, organizations will establish processes that provide for review of nonconformity by appropriate individuals in the organization. Such processes may have different levels of approval depending on the nature of the decision regarding the action to be taken for the nonconformity. A decision to "use as is," for example, may require engineering approval, since such a decision is effectively a change in design with liability implications, while manufacturing management may be permitted to approve a rework or scrap disposition.

Note that the second and third paragraphs of this clause require that all nonconforming products be addressed. If correction of the nonconformity is implemented, the organization must reverify the product to demonstrate conformity. There can be circumstances where organizations will not correct nonconforming product. Products that meet functional requirements are often used "as is," without taking action to make the product fully conform with all requirements, if such a decision will not affect the conformance of the end product ultimately delivered to a customer. Also, nonconforming product may be scrapped or regraded, and purchased material may be returned to a supplier. We recommend that when "use as is" is invoked, a change to less stringent requirements (i.e., possibly requiring design change action) should be considered.

For service, the clause attempts to recognize the reality that nonconforming service often cannot be controlled by the methods described. The organization can, perhaps, re-perform the service, but to segregate it is impossible! If a bank has a requirement that tellers be courteous to all customers, and a teller "insults" a customer, how can the "insult," the nonconforming service, be controlled? Like the fairy-tale genie that could not be stuffed back into the magic lamp, once the nonconforming service is delivered there is no way to control this nonconforming product. Thus, if the organization delivers services, it will need to think carefully about how to conform to this clause in a manner that adds value for both itself and its customers. Of course, in the service industry, "recovery" actions are common—the idea being to delight the customer with the organization's responsiveness to a difficult service situation.

Questions to Ask to Assess Conformity

- Is there a process to ensure that product that does not conform to requirements is identified and controlled to prevent unintended use or delivery?

- Is there evidence of appropriate action being taken when nonconforming process outputs, products, or services have been detected after delivery or use has started?

- Are appropriate concessions obtained from the customer when required?

- Is corrective action initiated in appropriate situations when nonconformities, including those discovered after delivery, are discovered?

- Is documented information retained as to the decisions made on nonconformities and who made those decisions?

Definitions (Refer to ISO 9000:2015)

- Concession
- Conformity
- Correction
- Corrective action
- Customer
- Defect
- Nonconformity
- Regrade

- Release

- Repair

- Review

- Rework

- Scrap

Considerations for Documented Information to Be Maintained and/or Retained

Clauses 8.5, 8.6, and 8.7 have no specific requirements for maintaining documented information on the processes involved. Clause 7.5.1 requires the organization to determine what documented information is necessary for the effectiveness of the QMS. Clause 8.5.1 requires availability of documented information to define product and service characteristics, activities to be performed, and results to be achieved.

We recommend that the documented information maintained include both procedure and work instruction information needed to produce the product or deliver the service. Consideration needs to be given to maintaining documented information for:

- Describing the interactions with other processes of the QMS, such as design and development

- Instructions used for product and service provision to the extent needed to ensure requirements are clearly understood

- Processes critical to preservation of product and service integrity

- Processes for ensuring appropriate measuring and testing resources, and processes for their use

- Processes for control of product identity and traceability where required

- The job titles of personnel responsible for release of product

- The process for control of nonconforming process outputs, products, and services, including such things as segregation, identification, and positive exclusion from unintended use

For clause 8.5—*Production and service provision*—documented information is required to be retained to comply with traceability requirements. Clause 8.5.6 also requires retention of documented information describing the results of reviews of changes, personnel authorizing the changes, and any necessary actions required from the review of changes. The organization should also consider retaining documented information for:

- Results of process control activities for significant product and service production and service provision

- Criteria for determining which processes require validation and results of validation and revalidation

- Property owned by the customer or an external provider that is incorrectly used, lost, damaged, or otherwise found to be unsuitable for use and reports of these conditions to the owner

- Completion of actions required for preservation of product

- Completion and conformity of post-delivery activities

For clause 8.6—*Release of products and services*—documented information that all acceptance criteria have been met is required to be retained before delivery. Documented information is also required on the person(s) authorizing release of products and services for delivery to the customer. Other aspects of release for which the organization should consider retaining documented information include:

- Objective evidence that all stages of product verification have been completed in accordance with planned arrangements

- Objective evidence of customer (and other party where required) approval for any product

Clause 8.7—*Control of nonconforming process outputs, products, and services*—requires retention of documented information on actions taken on nonconforming process outputs, products, and services. Documented information is also required of any concessions approved and the person or authority making the decisions related to the nonconformity. Other areas where the organization should consider retaining documented information include:

- Data related to the nonconformance and any related trends

- Decisions as to when to embark on corrective action

8

Clause 9—Performance Evaluation

CLAUSE 9.1—MONITORING, MEASUREMENT, ANALYSIS, AND EVALUATION

9.1.1 General

What Does the Clause Say?

Clause 9.1.1 requires the organization to determine:

- What needs to be monitored and measured

- The methods for monitoring, measurement, analysis, and evaluation, as applicable, to ensure valid results

- When the monitoring and measuring shall be performed

- When the results from monitoring and measurement shall be analyzed and evaluated

The organization shall also evaluate the performance and effectiveness of the QMS and retain appropriate documented information as evidence of the results of monitoring and measurement activities.

What Does the Clause Mean?

Although the requirements in this clause are similar to the requirements in clause 8.1 of ISO 9001:2008, several differences from the prior edition may require organizations to review existing methods and make adjustments. Examples of differences include:

- Methods for monitoring, measurement, analysis, and evaluation need to be determined

- A requirement for determining methods to ensure valid results versus the 2008 language of "determination of applicable methods"

- Addition of "retain appropriate documented information as evidence of the results"

- Evaluate the performance and effectiveness of the QMS versus the past requirement to "ensure conformity of the quality management system"

- Deletion of a requirement to use statistical techniques, which is now mentioned as a note to clause 9.1.3

These examples underscore the importance of careful review of the ISO 9001:2015 requirements to ensure processes address requirements and are effective.

9.1.2 Customer Satisfaction

What Does the Clause Say?

The organization shall:

- Monitor customer perceptions of the degree to which customer needs and expectations have been fulfilled

- Determine the methods for obtaining, monitoring, and reviewing this information

A note to this clause describes a few options for obtaining customer perceptions of the degree to which requirements have been met. As is always the case, the note is not normative (i.e., it does not state a requirement).

More interesting than what this clause requires is what it does not require. Customer satisfaction, one of the most fundamental organizational activities as a driver of improvement, continues to be very weak and wishy-washy. Our view is that it does not push organizations far enough to focus on addressing this activity.

What Does the Clause Mean?

As indicated earlier, the current requirements take a small step in requiring organizations to "listen" to their customers. Organization activities can range from very restricted (e.g., a customer satisfaction survey) to very robust. The wording of the requirements included in clause 9.1.2 is a compromise between requiring almost no effort to assess and understand customer satisfaction and compelling vigorous measures to understand the voice of the customer.

The only nuanced difference in 2015 is the requirement for the organization to determine the methods for obtaining information relating to customer needs and expectations versus requirements. Some may view the 2015 wording as a positive step. We believe more is needed to address this very important aspect of the QMS. If an organization believes that customer focus is a core value (one of the QMPs), then processes should be considered to assess its customer satisfaction performance. The nonnormative note provides activities that organizations can consider to make the processes robust.

Almost any credible process that is deployed will be acceptable to meet the minimal requirements of clause 9.1.2.

9.1.3 Analysis and Evaluation

What Does the Clause Say?

The organization is required to analyze and evaluate appropriate data and information arising from monitoring and measurement.

The output of the analysis shall be used to evaluate:

- Conformity of products and services to requirements

- Degree of customer satisfaction

- Performance and effectiveness of the QMS

- If planning has been effectively implemented

- Effectiveness of actions to address risks and opportunities

- The performance of external providers

- The need for improvements to the QMS

Analysis and evaluation of output is also required to be an input to the management review process (see clause 9.3).

What Does the Clause Mean?

Several tweaks were made to clause 9.1.3 that organizations will need to consider when assessing their processes for conformity to the 2015 edition of ISO 9001. Processes conforming to earlier versions of ISO 9001 may require modification. Requirements that may be subtle but important to review include analysis that is used to evaluate:

- The degree of customer satisfaction

- Conformity of products and services to requirements

- Performance and effectiveness of the QMS

- The effectiveness of planning

- The effectiveness of actions taken to address risks and opportunities

- The performance of external providers

- The need for improvements to the QMS

Consider the Potential Interactions as They Apply to Your QMS

Performance evaluation should be implemented in some form for all QMS processes. Therefore, clause 9.1 interacts with and should be considered for every element of the QMS.

Implementation Tips

One approach that has proven to be effective and efficient for addressing the requirements of clause 9.1.1 is to first ensure the key processes required for product realization, support, and improvement are identified. We recommend careful attention to clauses 8.1 and 8.5, which underscore the requirements for planning and controlling processes. For each key process, the inputs to and outputs from the process can be defined. Once the outputs of key processes are defined and understood, it is possible to determine how to measure or monitor the outputs to ensure that they meet requirements. While considering outputs and how to measure or monitor them, it is advisable to also consider and provide evidence of any documented information (i.e., records) that may be required to ensure conformity to requirements.

Such an approach can be implemented using a template consisting of a 3×2 matrix of blocks. The upper-left block can be used for listing the process inputs, the top middle block for listing the activity characteristics, and the upper-right block for listing the process outputs. The lower-left block can contain the interfaces or interactions with the activity being analyzed, the lower middle block for listing the records required, and the lower-right block for listing the metrics to be used to ensure that the output meets requirements.

A simplified example of such a template completed for the corrective action process (see 10.2) could look like Figure 8.1.

The planning and implementing of monitoring, measurement, analysis, and evaluation processes can be accomplished by any means the organization chooses. We have found that having a model and a defined approach for addressing these requirements ensures appropriate attention to the planning of the monitoring, measurement, analysis, and evaluation processes. The completion of a matrix as described earlier for

Inputs	Process	Outputs
• Undesirable process or condition • Customer complaint	• Review request • Proceed with corrective action (CA) • If yes, develop CA plan • Implement plan • Check effectiveness	• Implemented CA • Documentation to management review file

Interactions	Records	Metrics
• Depends on CA; could be purchasing department, design engineering, training department, etc.	• CA plan • CA effectiveness	• CA log

Figure 8.1   Process characteristics.

key processes is one approach to ensure we are addressing the requirements of this clause.

The requirements in clause 9.1.2 are very general. So what can an organization do to conform in a way that adds value? Since it is the organization's responsibility to decide what customer perception information it will monitor (and, where appropriate, measure), a few examples of possibilities are provided here for consideration.

Organizations that function in regulated markets may choose to monitor customer reports of product deficiencies. A large automobile manufacturer might measure customer satisfaction as reflected in surveys mailed to new car owners. Service providers may choose to use focus groups to probe customers' perceptions. Software suppliers could monitor reported bugs from field installations. The important point is that the organization decides what to monitor and what methods to use.

Clause 9.1.1 mentions monitor and measure. Clause 9.1.2 mentions monitor. Monitoring may provide less information than measuring. On the other hand, measurement may be used as part of a monitoring process, or the results of monitoring may indicate a need to gather more information through measuring. An organization might monitor or measure or do both depending on the data it requires to assess process performance.

Examples of sources of customer satisfaction information that could be used to meet the requirements clause 9.1.2 include the following:

- Customer complaints

- Returns

- Warranty information

- Customer-satisfaction studies

- Results from focus group meetings

- Customer tracking studies

- Questionnaires and surveys

- Reports from consumer organizations

- Direct customer communication

- Benchmarking data

- Industry group information

- Trade association information

In most organizations, many sources of information about customer satisfaction and dissatisfaction are available. This information is often poorly organized and even more poorly used. In many cases organizations do not recognize the value of the data they already have. This requirement of ISO 9001:2015 should encourage organizations to better use the gold mine of information already available to them.

The organization also must decide the extent to which its processes go beyond mere conformance to requirements to meet the unstated needs and expectations of customers. This should include price and delivery considerations. It is up to the organization to decide how far to go in this direction, and the decision should be related to the organization's quality policy and objectives.

ISO 9001:2015 takes one additional step in the area of customer satisfaction—it includes a requirement that the methods for obtaining and using customer information must be determined. This means that the organization must think about how to gather information and what will be done with it after it is gathered. Many organizations gather information—some make an effort to understand what the information means, but few take real actions to improve the organization and its processes based on data that indicate the degree to which customer needs and expectations have been fulfilled. The intent of ISO 9001:2015 is to encourage organizations to plan what to gather, to gather it, to analyze and understand it, and to take appropriate action. It is meaningless if the quality policy of an organization pronounces a high regard for customer satisfaction but does not deploy processes to assess satisfaction levels and does not take actions to improve performance.

Effective conformity to the requirements of clause 9.1.3 requires an organization to be dedicated to improvement. That kind of organization will view the requirements of all elements of the QMS as linked in the sense that the organization should function on a closed-loop basis. This means:

- Continual measuring of processes and products

- Analyzing and evaluating data

- Using the facts derived from analysis to improve both the QMS and the products and services of the organization

The gathering, analysis, and evaluation of data are powerful tools to drive continual improvement. The organization should give serious consideration to documenting expectations for these activities at least in the areas related to:

- Customer satisfaction (see 9.1.2)

- Conformity to product requirements (see 8.2.4)

- Characteristics and trends of processes and products, including risk assessment actions (see 6.1.1) and suppliers

Information from analysis and evaluation of data is also a valuable input to the management review process. Analysis may also be conducted as part of the management review itself. Clause 9.1.3c requires analysis and evaluation to determine the suitability and effectiveness of the QMS, and clause 9.3 requires that top management review the QMS to ensure its suitability, adequacy, and effectiveness. Top management has a good deal of flexibility in conforming to these requirements. A few of the options for top management include the following:

- Data may be analyzed and evaluated off-line, and information from the analysis may be provided as input to top management for use in determining suitability and effectiveness of the system. This may be typical of larger organizations or organizations with dedicated analytical staff.

- Data may be provided to top management, who could direct an analysis as an input to management review. This may be more typical of a small organization.

The organization is required to analyze and evaluate data to identify areas where opportunities for improvement exist. The analysis and evaluation must provide information in several specific areas (see 9.1.3a–g): (a) customer satisfaction and/or dissatisfaction; (b) conformance to customer requirements; (c) characteristics of processes, product, and their trends; and

(d) external providers. The specific information that is appropriate may differ based on the type and size of the organization as well as on the product category.

Customer satisfaction and/or dissatisfaction information is different from the information related to meeting customer requirements. It is possible for customers to be satisfied with product that is nonconforming or to be highly dissatisfied with product that is in full conformance with requirements. In either case, identification of the situation offers an opportunity to change requirements to reflect actual customer needs. Measurement of customer information (as covered in clause 9.1.2) may include such items as issues of importance to customers, gaps in meeting customer expectations, customers' desires for changes in characteristics or features of the product, the relative satisfaction of customers with the organization and its competitors, and the organization's most significant customer-complaint areas. The appropriate information on customer satisfaction and/or dissatisfaction may depend on the nature of an organization's relationships with its customers. For example, large organizations selling products to consumers through multiple distribution channels may need information related to several tiers of customers in their value chain. This may include information on the several distribution channels and on consumers as well. This could be the case for toy manufacturers (hardware), home computer software providers (software), and airlines (service). At another extreme, organizations with a single customer and day-to-day personal customer contact may have very different information needs.

Information related to conformance to customer requirements describes how well the customer's requirements (stated and unstated) are being met. It includes information related to requirements as well as needs and expectations not stated by the customer. Customer inspections/tests, field problem reports, and warranty returns, as well as informal complaints communicated via e-mail or telephone calls or other channels (e.g., salesmen), are typical of customer information related to conformance to requirements that should be considered for analysis.

Information on characteristics of processes, products, and their trends can be derived from analysis of product and process data obtained from the measurement process. This may also include information from aggregation and analysis of both internal operational data and feedback from customers.

Information on suppliers can be developed from analysis of supplier performance data. It may include information on both high- and low-performing suppliers. Since purchased material is often a significant percentage of total cost of goods sold, it makes business sense to invest appropriately in ensuring excellent performance by suppliers (e.g., part per million quality levels or 100% on-time delivery).

Sales personnel are often a good source of input data related to customer satisfaction and product conformity to customer requirements and expectations.

For hardware and processed materials, information related to conformance to customer requirements that should be considered for analysis may include such items as the most numerous/significant nonconformities reported by the customer, costs of customer returns, and significant design changes resulting from customer feedback. Information on characteristics of processes, products, and their trends may include manufacturing process capabilities, types of significant assembly defects, order entry error rates, cost of quality data, and statistical process control data. It may also include such items as line balance information, cell cycle times, and other information needed to improve internal processes such as scheduling and cycle time.

For services, information related to conformance to customer requirements that may be considered for analysis includes such items as nonconformities related to service delivery by customer-contact employees and/or service deliverers. The information may also include causes of late service performance, the most significant reasons for service outages, unavailability of service due to overcapacity scheduling, inadequate documentation, billing, and other accounting errors.

Information on characteristics of processes, product, and their trends may include significant causes of process backlogs, ability of key service processes to deliver required services when requested by the customer, time needed to respond to service requests, satisfaction with delivered training, acceptability of consulting service, late delivery of service, and late and/or over-budget development projects. Any and perhaps all of these causes of service delivery nonconformity or customer dissatisfaction can be considered for analysis and evaluation.

In the software area, the organization should consider gathering and analyzing information related to the most numerous/ significant nonconformities reported by the customer, the cost to correct the nonconformity, and issues related to installation, start-up, and integration of software modules. Since the majority of the cost of ownership of software occurs after purchase or product release, gathering and analysis of nonconformity costs can prolong the lifetime of software products. Alternatively, not addressing life-cycle costs can render software products untenable.

Information on characteristics of processes, products, services, and their trends may include rate of decline of bugs found, on-time release, acceptability of design reviews, and controlling change.

Collection of data without developing the data into useful information is a waste of organizational resources. The purpose of analysis and evaluation is to convert data into usable information. One of the most important considerations in establishing data-collection methods is to determine how the data will be used. When a data collection scheme has been well designed, the analysis and evaluation effort is simplified. Poorly designed data-collection systems not only are inefficient but also can yield misleading information and add a lot of non-value-added time.

CLAUSE 9.2—INTERNAL AUDIT

9.2.1 Untitled

What Does the Clause Say?

Conduct internal audits at planned intervals to provide information on whether the QMS:

- Conforms to the organization's own requirements for its QMS

- Conforms to the requirements of this International Standard

- Is effectively implemented and maintained

What Does the Clause Mean?

The internal audit process is grouped with clauses on performance evaluation. Internal auditing is a form of performance evaluation focused on assessing the conformity of the QMS and its processes to requirements. It should be deployed to assess compliance of processes, to drive improvement of product realization, and to enhance the ability of the organization to meet customer expectations. This clause establishes clear objectives of and requirements for the internal audit process.

9.2.2 Untitled

What Does the Clause Say?

The internal audit process of the organization shall:

- Plan, establish, implement, and maintain an internal audit program

- Define the criteria and scope for each audit

- Select auditors based on competence criteria

- Ensure audits are conducted in an impartial and objective manner

- Ensure results of audits are reported to relevant management

- Take appropriate correction and corrective actions without undue delay

- Retain documented information of audit results and audit program implementation

The note to clause 9.2.2 references ISO 19011, which is titled *Guidelines for auditing management systems.* This International Standard is a useful reference source.

What Does the Clause Mean?

The essential elements of an internal audit process are contained in the requirements stated in clause 9.2.2. The requirements stated are similar to the previous requirements, although a few of the previous requirements have been dropped (e.g., the requirement to determine conformity with "planned arrangements"). Some text has been tweaked to improve clarity, and a few new requirements have been added (e.g., the internal audit program shall be "established, implemented, and maintained"). We believe most organizations that have implemented a conforming process will have little difficulty making minor "improvements" to address the current requirements.

Consider the Potential Interactions as They Apply to Your QMS

Although the audit program required by clause 9.2 interacts with all processes of the QMS (and hence all clauses of ISO 9001:2015), its development and deployment has the greatest interaction with the processes of the QMS related to:

- Clause 5 Leadership

- Clause 9.3 Management review

- Clause 10.1 Corrective action

- Clause 6 Risk-based thinking

Implementation Tips

Internal audit of the QMS as a form of measurement continues to be an essential process to provide confidence in the effective implementation of the QMS. To better understand the role of internal audit, it is useful to consider its role as complementary to that of two other forms of QMS evaluation—management review and self-assessment.

The evaluation of a QMS can vary in scope and encompasses the following three major approaches:

- Audit, the subject of this section

- Review of the QMS, addressed in the next clause (9.3)

- Self-assessment, addressed in ISO 9004 in the text and in an annex, and in an ASQ technical report

Audits are used to evaluate the adequacy of QMS documentation, conformance to its requirements, and the effectiveness of implementation. The results of audits can be used to identify opportunities for improvement. ISO 9001:2015 continues to require auditors to determine the effectiveness of implementation. Determination of overall QMS suitability and effectiveness is left to top management, who uses audit results and other data as inputs to make that evaluation.

In auditing language, internal audits are considered first-party audits—audits conducted by or on behalf of the organization for internal purposes, which can form the basis for an organization's self-declaration of conformity. Second-party audits are conducted by customers of the organization or by other persons on behalf of a customer. Third-party audits are conducted by external independent audit service organizations (e.g., a registrar). Such organizations can verify conformity with requirements such as those of ISO 9001:2015.

One role of top management is to ensure the organization carries out regular, systematic evaluations of the suitability, adequacy, effectiveness, and efficiency of the QMS with respect to the quality policy and objectives (see 5.1.1). This systematic evaluation (see ISO 9001:2015, clause 9.3) can include consideration of the need to adapt the quality policy and objectives in response to the changing needs and expectations of interested parties (see 4.2). The review includes determination of the need for actions. Among other sources of information, audit reports are used for management review of the QMS.

An organization's self-assessment is a comprehensive and planned review of its activities, and the results are referenced against the QMS or a model of excellence. The use of self-assessment methodology can provide an overall view of the performance of the organization and the degree of maturity of the QMS. It can also help identify areas requiring improvement in the organization and determine priorities. It is normal for such self-assessments to go well beyond assessment of conformance to requirements. Self-assessments can look for opportunities for the organization to improve its effectiveness, efficiency (not a requirement of ISO 9001:2015), and performance and often attempt to identify best practices that may be portable to other areas of the organization. Although not an ISO 9001:2015 requirement, a robust self-assessment process is a valuable companion to the internal audit process of an organization. ASQ Z1 TR1-2012—*Guidelines for performing a self-assessment of a quality management system*—is a useful reference for conducting self-assessments.

ISO 9001:2015 requires internal quality audits to be conducted periodically. These audits should be used to determine conformity to the requirements of ISO 9001:2015 and the degree to which the QMS has been effectively implemented and maintained. An indicator of problems with the effectiveness of the QMS is the occurrence of high numbers of customer complaints or of high levels of scrap and rework within the organization. As stated in clause 9.1.2, organizations are expected to monitor customers' perceptions. Internal auditors often use this information to identify product realization processes that require further investigation regarding the extent to which they have been effectively implemented and maintained. In a similar fashion, scrap and rework information may be of value for identifying subject areas for internal audits.

Whatever factors and methods are used, internal quality audits may be performed on processes of the QMS or on the entire system. Whatever approach is used, details need to be established in plans and the organization should determine priorities for the processes to be audited. Previous audit results should be used in developing the prioritization. The scope of each audit should be clear, and the frequency of audits within the audit program and the audit methodology should be identified.

One aspect of internal auditing that is often underaddressed is the competence of internal auditors. Just because an individual is a subject matter expert does not mean that he or she will be a competent internal auditor. Internal auditors should be qualified as auditors. This is particularly necessary for "guest" auditors or technical experts from other functions who are used to provide product or process technical expertise to the effort to evaluate effectiveness but who tend to be inexperienced at auditing. As is true about all work that is performed by an organization, those who perform the work are required to be competent. One approach to ensuring competence of internal auditors is to establish minimum requirements in terms of training and/or experience and/or working with experienced auditors until competence has been demonstrated to a member of management by some method or combination of methods.

Audit results should be documented in a written report, and the documented information (i.e., the audit report) should indicate nonconformities and deficiencies. It may also include opportunities for improvement of conforming processes and best practices that are observed that can be considered for deployment elsewhere in the organization. Target dates should be established for responding to audit findings, and organizations should take corrective action in a manner and with timing that is appropriate for the circumstances. Audit results are required to be inputs to management reviews.

The audit process also needs to ensure that documented information is retained as evidence of the implementation of the audit program and the audit results.

CLAUSE 9.3—MANAGEMENT REVIEW

9.3.1 General

What Does the Clause Say?

The organization's QMS shall be reviewed by top management, at planned intervals, to ensure its continuing suitability, adequacy, and effectiveness. This review shall be planned and carried out taking into consideration:

- The status of actions from previous management reviews

- Changes in external and internal issues that are relevant to the QMS, including its strategic direction

- Information on the quality performance, including trends and indicators for:

  — Nonconformities and corrective actions

  — Monitoring and measurement results

  — Audit results

  — Customer satisfaction

  — Issues concerning external providers and other relevant interested parties

  — Adequacy of resources required for maintaining an effective QMS

  — Process performance and conformity of products and services

- The effectiveness of actions taken to address risks and opportunities (see clause 6.1)

- New potential opportunities for continual improvement

What Does the Clause Mean?

Management review of the quality system is the responsibility of top management. This is no different from the previous requirement for management review. The current edition retains all previous requirements except for the reference to preventive action, which is addressed when considering risks and opportunities elsewhere in the current edition (see clause 6.1). It also characterizes recommendations for improvement as new potential opportunities for continual improvement, and input related to customer feedback as customer satisfaction input.

So, although some of the words have been tweaked, the planning and implementation of management review processes has remained almost the same as was required by ISO 9001:2008.

Implementation Tips

The notion of management review is to stand back, look at the effectiveness of the QMS, examine performance, and decide what changes are needed to further improve the QMS.

Many inputs to management review can be listed beyond those required by ISO 9001:2015. At the top of the list might be progress on measures related to meeting the quality objectives.

A few characteristics of successful management reviews are things like:

- A top management *attitude* that the system can and should be improved. The objective of management review is to determine how. Management reviews are not celebrations of what went right; they are reviews to determine how to make more things go right—and better.

- *Preparation!* It is essential. Someone needs to get all the inputs together and make sense of them before the review starts. This is so top managers can see issues from all relevant perspectives. For example, aggregation of internal process data with customer complaint information can paint a clear picture of an opportunity to improve customer satisfaction while the improvement process also reduces costs.

- *Follow-up* is necessary to make certain that decisions are clear and action is taken. Frequent and formal progress reviews are needed. Robust, effective management reviews can be developed by starting them at the beginning of QMS implementation. Starting the process early can facilitate systematic improvement of the review process over the implementation phase so that it becomes a key element of organizational success.

- *Having and following a crisp agenda.* The reviews are not problem-solving meetings but rather an opportunity to assess status and make decisions regarding who is going to do what and by when.

9.3.2 Management Review Inputs

What Does the Clause Say?

Inputs for management reviews are required to include the status of actions from previous management reviews and information on the performance and effectiveness of the QMS. Other areas that shall be considered include the adequacy of resources, the effectiveness of actions to address risks and opportunities, and opportunities for improvement. The process shall also consider trends in:

- Customer satisfaction

- Feedback from interested parties

- The extent to which quality objectives have been met

- Process performance and conformity of products and services

- Nonconformities and corrective actions

- Monitoring and measurement results

- Audit results

- The performance of external providers

What Does the Clause Mean?

To perform an effective management review requires input regarding the activities articulated in clause 9.3.2 to guide management decision making, setting priorities, and allocating resources. The effectiveness of the management review process is dependent on the information provided. This clause identifies a minimum set of inputs for the process to be considered during the review. Not every activity may be discussed in a review, but management should at least consider all the activities listed.

Implementation Tips

The inputs that are required for consideration in clause 9.2 should not be considered as a complete list. The organization should decide what needs to be covered. Management, for example, may require a periodic report on quality objectives. It may be effective to establish a process that identifies who will develop what inputs and who will be the presenter (or question answerer) during management reviews.

9.3.3 Management Review Outputs

What Does the Clause Say?

Outputs of the management review shall include decisions and actions related to:

- Identification of opportunities to improve the QMS

- Making changes to the QMS that have been identified or are anticipated

- Resources

Documented information is required to be retained as evidence of the results of management reviews.

What Does the Clause Mean?

As was indicated earlier, key words in the requirements are *decisions* and *actions*. After considering the inputs, management has the capability to initiate actions and allocate resources to improve the effectiveness of the QMS.

Clause 9.3.3 is very clear and unambiguous. It means what it says.

Consider the Potential Interactions as They Apply to Your QMS

The management review clause has primary interaction with:

* Clause 5.1 Leadership and commitment

* Clause 8 Operation

* Clause 10 Improvement

Implementation Tips

See "Implementation Tips" under 9.3.1.

When deploying the management review process, it is recommended that the reviews be "orchestrated" as opportunities to address what requires attention and to allocate the priorities and resources to attack the issues—and not sessions to analyze and develop courses of action to resolve issues.

Earlier we recommended using the language of management (i.e., money—in case you have forgotten) when discussing issues with management, especially top management; recommendation is also made here for discussing issues during management review sessions in money terms.

When identifying opportunities to improve the QMS, its processes, and the products and services provided to customers, it is important to keep a focus on customer needs. There is no requirement to improve the product beyond the point where all customer requirements are met, but remember that top managers desire to ensure that customers' future needs are determined and met in a way that enhances customer satisfaction, thereby creating value. It is therefore important to recognize that customer needs and expectations can change often; thus, the management review is an opportunity to identify new customer requirements and establish actions to meet them, which is one often overlooked dimension of changes to the QMS that should be considered.

Questions to Ask to Assess Conformity

- Is evidence available to demonstrate that the organization has defined, planned, and implemented the monitoring and measurement activities needed to ensure conformity?

- Is evidence available to demonstrate that the organization has determined the need for and use of applicable methodologies?

- Has the organization determined when to analyze and evaluate the results from monitoring and measurement?

- Has the organization evaluated the quality performance and effectiveness of the QMS?

- Is the organization monitoring customer perceptions of the degree to which requirements have been met?

- Have methods for obtaining and using customer information been determined?

- Has the organization obtained information relating to customer views and opinions of the organization and its products and services?

- Has the organization determined the appropriate data to be collected, analyzed, and evaluated?

- Does the organization analyze and evaluate the appropriate data to determine the suitability and effectiveness of the QMS?

- Does the organization analyze appropriate data to identify improvements that can be made?

- Does the organization analyze appropriate data to provide information on customer satisfaction and/or dissatisfaction?

- Does the organization analyze appropriate data to provide information on conformance to customer requirements?

- Does the organization analyze appropriate data to provide information on characteristics of processes, product, and their trends?

- Does the organization analyze appropriate data to provide information on suppliers?

- Does the organization conduct periodic audits at planned intervals to provide information on whether the QMS conforms to the organization's own requirements?

- Do the periodic audits evaluate the conformity of the QMS to the requirements of ISO 9001:2015?

- Do the periodic audits evaluate the degree to which the QMS has been effectively implemented and maintained?

- Has the organization planned, established, implemented, and maintained an audit program (including the frequency, methods, responsibilities, planning requirements, and reporting) that considers the quality objectives, the importance of the processes concerned, customer feedback, changes impacting on the organization, and the results of previous audits?

- Has the organization planned the audit program taking into consideration the results of previous audits?

- Are the audit scope, frequency, and methodologies defined?

- Do the audit process and auditor assignment ensure objectivity and impartiality?

- Is there a process for reporting results to relevant management and maintaining records?

- Are necessary correction and corrective actions taken without undue delay?

- Does top management review the QMS at planned intervals to ensure its continuing suitability, adequacy, and effectiveness?

- Are management reviews planned and carried out taking into consideration the status of actions from previous management reviews?

- Are management reviews planned and carried out taking into consideration changes in external and internal issues that are relevant to the QMS, including its strategic direction?

- Are management reviews planned and carried out taking into consideration information on quality performance, including trends and indicators for nonconformities and corrective actions, monitoring and measurement results, audit results, customer satisfaction, issues concerning external providers and other relevant interested parties, adequacy of resources required for maintaining an effective QMS, and process performance and conformity of products and services?

- Are management reviews planned and carried out taking into consideration the effectiveness of actions taken to address risks and opportunities (see clause 6.1)?

- Are management reviews planned and carried out taking into consideration new potential opportunities for continual improvement?

- Does the output of the management reviews include decisions and actions related to any need for changes to the QMS, including resource needs?

- Does the output of the management reviews include decisions and actions related to continual improvement opportunities?

- Does management review input include results of audits, customer feedback, process performance, product conformity, status of preventive and corrective actions, follow-up actions from earlier management reviews, changes that could affect the QMS, and recommendations for improvement?

- Does the organization retain documented information as evidence of the results of management reviews?

Definitions (Refer to ISO 9000:2015)

- Audit

- Audit conclusion

- Audit criteria

- Audit findings

- Audit plan

- Audit program

- Audit scope

- Auditor

- Conformity

- Continual improvement

- Customer satisfaction

- Documented information

- External supplier

- Improvement
- Interested party
- Measurement
- Performance
- Process
- QMS
- Risk

Considerations for Documented Information to Be Maintained and/or Retained

- The organization is required to retain appropriate documented information as evidence of the results of implementation of monitoring and measurement activities that are conducted in accordance with the determined requirements

- The organization should consider documented information to define and deploy processes to determine what needs to be monitored and measured and the methods for monitoring, measurement, analysis, and evaluation, as applicable, to ensure valid results

- The organization should consider documented information to determine when the monitoring and measuring shall be performed

- The organization should consider documented information to determine when the results from monitoring and measurement shall be analyzed and evaluated

- The organization should consider documented information to monitor customer perceptions of the degree to which requirements have been met

- The organization should consider documented information to determine how it will obtain information relating to customer views and opinions of the organization and its products and services

- The organization should consider documented information to describe its processes for analyzing and evaluating appropriate data and information arising from monitoring, measuring, and other sources

- The organization should consider documented information to describe its processes for using the output of analysis and evaluation

- Documented information is required to be retained as evidence of the results of management reviews

- The organization should consider maintaining documented information to define all the requirements for the management review process, including inputs, activities, outputs, follow-up, and the duties and responsibilities of the participants

9

Clause 10—Improvement

It is not necessary to change. Survival is not mandatory.

—W. Edwards Deming

If'n you be too stoopid to git better, y'all ain't a gonna make it in the 'shine business. Dem cus'mers demand great hooch.

—Anonymous Tennessee moonshiner

CLAUSE 10.1—GENERAL

What Does the Clause Say?

Clause 10.1 requires organizations to determine and select opportunities for improvement and implement necessary actions to meet customer requirements and enhance customer satisfaction.

The opportunities for improvement stated for inclusion, as appropriate, are:

• Correcting, preventing, or reducing undesired effects

• Improving products and services to meet requirements as well as to address future needs and expectations

• Improving QMS performance and effectiveness

A note to this clause indicates that improvement initiatives can be reactive (e.g., corrective action), incremental (e.g., continual improvement), by step change (e.g., breakthrough), creative (e.g., innovation), or by reorganization (e.g., transformation). The note also includes the notion of "correction," which involves returning a nonconformity to its intended state and in most circles is not considered an improvement. You can, of course, argue that correcting the immediate problem may improve the current situation.

What Does the Clause Mean?

This clause requires ensuring an improvement mentality is institutionalized.

ISO 9001:2015 makes it clear that improvement opportunities shall be determined and selected. The key elements for improvement listed in 10.1 are improving the QMS results, improving products and services to meet known and predicted requirements, and improving processes to prevent nonconformities.

It should be apparent that if an organization invokes robust processes for establishing a quality policy, setting quality objectives, and analyzing data, and pursues corrective action, preventive action, and internal audit and management review, as is required by other clauses of this International Standard, it would be difficult not to improve.

Implementation Tips

When considering the development and deployment of processes to address the improvement requirements of ISO 9001:2015, it is important to recognize that improvement is critical to the sustainability of the organization. If the organization is not improving both the products and services it provides to its customers and the effectiveness (and the efficiency as well) of the processes employed to realize those products and services, its long-term survival is threatened. Competitors will not be standing still, and expectations of customers are likely to become higher. In the public sector, potential privatization of products and services offered is an incentive to emphasize improvement initiatives.

Clause 10.1 requires the organization to improve its effectiveness through improving its QMS results. What actions are necessary to consider to meet customer requirements and enhance customer satisfaction? To address this requirement the organization should consider aggressive improvements in *every* element of its QMS. It is most critical to understand, address, and sharpen process and element interactions. Any aspect, process, or element of the QMS that is not important to the organization's success should be identified and targeted for elimination or combination with other elements. There is an "underrecognition" of the interrelationships of processes in organizations. It is not uncommon, for example, to take corrective action in one area but not recognize and consider the issues that can potentially be created in other areas. To paraphrase the words of Peter Senge, the system can and often does push back. When implementing improvement projects, care must be taken to consider process interactions.

It is not only necessary to consider all aspects of the QMS but vital to do so. An attitude of improvement needs to permeate the culture and the behavior of the entire organization if the organization wants to survive. It may be sufficient in some people's minds to write a simple corrective action procedure and have the quality activity create evidence that such a procedure has been followed a few times. Such an approach may get or keep an organization certified, but may well result in long-term negative business results and is not sufficient to ensure the survival of the organization. The intent of clause 10.1 is to achieve an integrated improvement mentality throughout the organization. This means using as many avenues as possible and ensuring the intention is clear to all in the organization that improvement is a core value. The ideal is to achieve a state where everyone in the organization is, and feels a need to be, a contributor to continual improvement.

When a robust improvement process has been developed and deployed, it can be helpful to ask questions to ensure that the organization is meeting at least the minimum requirements of ISO 9001:2015. A much broader set of questions can be developed for self-assessment that will highlight the degree of maturity of the improvement process. Questions can range from conformity with the minimum requirements in this International Standard to having an improvement process in place that is making ongoing and significant contributions to improving organizational performance and progressing the organization on the path to world-class performance levels.

It also may be helpful to refer to QMP 5 in ISO 9000:2015 to review the essence of improvement as a foundation of ISO 9001:2015.

CLAUSE 10.2—NONCONFORMITY AND CORRECTIVE ACTION

10.2.1 Untitled Clause

What Does the Clause Say?

When nonconformities occur, including those arising from complaints, the organization shall react to each nonconformity and, as applicable, take action to control and correct it and to deal with the consequences.

The organization is required to evaluate the need for action to eliminate the cause(s) of the nonconformity to ensure that it does not recur or occur elsewhere by reviewing the nonconformity, determining its causes, and determining whether similar nonconformities exist, or if conditions exist under which similar nonconformities may occur.

The organization is also required to implement any action that is needed, to review the effectiveness of any corrective action taken, and to make changes to the QMS as appropriate. Corrective actions, it is noted, shall be appropriate to the effects of the nonconformities that are encountered.

What Does the Clause Mean?

The requirement for organizations to react to nonconformities and to implement corrective action has been part of ISO 9001 from the beginning. It involves taking action to eliminate the causes of nonconformities. There is little new in the requirements for addressing the handling of nonconformities for organizations that have corrective action processes (CAPs) in place that conform to ISO 9001:2008.

Nonconformities need to be evaluated, and the causes of their occurrence need to be determined. Evaluations of the nonconformities should indicate what corrective actions to take to eliminate the causes of the nonconformities.

Once actions to correct the cause(s) of the nonconformity have been determined, they need to be implemented and their effectiveness reviewed; changes to the QMS, if any, should be implemented as appropriate. Many organizations meet this requirement by having someone review the action taken and then sign off that the action taken is effective.

Implementation Tips

Addressing and resolving nonconformity is best achieved by having documented information (i.e., a procedure) that details how the organization shall react when nonconformity occurs. The process needs to address several activities, such as:

- How to deal with the consequences of the nonconformity
- How to initiate action to control the nonconformity
- Determining the causes of the nonconformity

- How to decide on a course of action
- Determining whether similar nonconformity can exist elsewhere
- Initiating any actions needed
- Reviewing the effectiveness of the actions taken
- Making any changes to the QMS

Deploying such a process will ensure consistent handling of nonconformity. It is also important to ensure that everyone in the organization understands that there is a distinct difference between correction and corrective action.

10.2.2 Untitled Clause

What Does the Clause Say?

The organization shall retain documented information as evidence of the nature of the nonconformities and any subsequent actions taken and the results of any corrective actions.

What Does the Clause Mean?

Clause 10.2.2 is very short and explicit. The organization needs to retain records about nonconformities, the actions taken, and the results of any corrective action.

Implementation Tips

Overall, the contents of 10.2.1 and 10.2.2, when considered together, are very similar to previous requirements for corrective action.

It is advisable to have a process defined and documented (i.e., a procedure) for addressing nonconformity, correction of nonconformity, and corrective action. It is an unfortunate fact that in many organizations the difference between correction and corrective action is not understood or is ignored. Correction of problems is common. Corrective action is not so common. So the challenge to organizations is to develop a process that requires true corrective action (i.e., identification and either elimination or at least minimization of root causes of nonconformity). A bigger challenge is to ensure that all personnel understand the meaning of corrective action and are competent to follow the process.

It is also advisable to include in a procedure that, in some cases, action may be neither required nor appropriate. If the nonconformity is minor and an isolated condition, the risks or costs associated with taking corrective action may not be justified. Deciding not to take corrective action as a management decision is acceptable, as long as it is a conscious decision that considers the circumstances and risks and, if and where appropriate, is discussed with customers.

CLAUSE 10.3—CONTINUAL IMPROVEMENT

What Does the Clause Say?

The organization shall continually improve the suitability, adequacy, and effectiveness of the QMS.

The organization shall consider the outputs of analysis and evaluation, and the outputs from management review, to decide whether areas of underperformance or opportunities shall be addressed as part of continual improvement.

What Does the Clause Mean?

Improvement of the QMS continues to be emphasized. Prior requirements to improve the QMS through the use of the quality policy, quality objectives, audit results, analysis of data, corrective and preventive actions, and management review are no longer explicitly present. The requirement to continually improve the performance and effectiveness of the QMS, however, is embodied when Clause 10 is considered in its entirety.

Implementation Tips

The best implementation tip to continually improve the suitability, adequacy, and effectiveness of the QMS is to develop, document, and deploy a robust improvement process. A weak variation of a corrective action process will not address the intent of this requirement. Our experience is that many organizations do not take improvement seriously, or they characterize corrective actions as improvement to have some objective evidence of having addressed risk aversion.

Some specific items to consider including in an improvement process are as follows:

- Performing a failure mode, effects, and criticality analysis (FMECA) in the design stage

- Performing reliability analyses to make reliability tradeoffs (e.g., mean time between failures, Weibull modeling)

- Fault tree analysis

- Data analysis of other areas of the organization to identify problems that may become problems elsewhere

- Corrective actions to determine whether there is potential to expand the scope to embrace improvement and risk aversion

- Consideration of analytics employed in Six Sigma, lean, and other similar improvement methodologies

It can also be helpful to train individuals on the use of contemporary tools (either computer or pencil-and-paper based) to investigate the causes of underperformance, to support improvement, and to develop improved methodology. In prior editions such activities were called "the use of statistical methods."

CONSIDER THE POTENTIAL INTERACTIONS AS THEY APPLY TO YOUR QMS

All the contents of the improvement clause interact with every element of ISO 9001:2015. Improvement initiatives should be considered for every QMS process at every level in the organization with objectives established where and when appropriate.

QUESTIONS TO ASK TO ASSESS CONFORMITY

- How does the organization determine and select opportunities for the continual improvement of the QMS?

- How does the organization implement necessary actions to meet customer requirements and enhance customer satisfaction?

- Does the organization consider selection of opportunities to improve processes to prevent nonconformities?

- Does the organization consider selection of opportunities to improve products and services to meet known and predicted requirements?

- Does the organization consider selection of opportunities to improve QMS results?

- Does the organization take corrective action to control and correct nonconformities?

- How does the organization address the consequences of nonconformity?

- Does the organization evaluate the need for action to eliminate the cause(s) of the nonconformity in order to ensure that it does not recur?

- Does the organization consider whether similar nonconformities exist or could potentially occur?

- How does the organization implement any action needed?

- How does the organization review the effectiveness of any corrective action taken?

- How does the organization make changes to the QMS if necessary?

- Does the organization retain documented information as evidence of the nature of the nonconformities, any subsequent actions taken, and the results of any corrective action?

- Is the corrective action taken appropriate to the impact of the problems encountered?

- How is the organization continually improving the suitability, adequacy, and effectiveness of the QMS?

- How is the organization considering the outputs of analysis and evaluation, and the outputs from management review to determine whether there are areas of underperformance or opportunities to address as part of continual improvement?

- How is the organization selecting and utilizing applicable tools and methodologies for investigating the causes of underperformance and for supporting continual improvement?

DEFINITIONS (REFER TO ISO 9000:2015)

- Complaint
- Continual improvement
- Correction

- Corrective action

- Customer

- Customer satisfaction

- Documented information

- Improvement

- Nonconformity

- Organization

- Output

- Product

- QMS

- Requirement

- Service

CONSIDERATIONS FOR DOCUMENTED INFORMATION TO BE MAINTAINED AND/OR RETAINED

- We recommend documented information (i.e., a procedure) that defines requirements for determining and selecting opportunities for improvement

- We recommend documented information (i.e., procedures) for implementing improvement opportunities, as appropriate, for improving processes to prevent nonconformities, for improving products and services to meet known and predicted requirements, and for improving QMS results

- We recommend documented information (i.e., a procedure) that ensures improvement objectives are considered at all levels and for all roles in the organization

- We recommend documented information (i.e., procedures) that addresses requirements for all phases of handling nonconforming materials

- We recommend documented information (i.e., procedures) that addresses requirements for all phases of corrective action processes

- Documented information is required to provide evidence of nonconformities and the effectiveness of any action to address the nonconformity and the results of corrective actions

- We recommend documented information (i.e., a procedure) that ensures that the difference between correction and corrective action is understood

- We recommend training all staff, as appropriate, on the difference between correction and corrective action and how to approach and conduct a process improvement project

10

Annexes and Bibliography

It is only the unimaginative who ever invents. The true artist is known by the use he makes of what he annexes, and he annexes everything.

—OSCAR WILDE

ISO 9001:2015 contains two annexes and a bibliography. The annexes are informative, which means that they do not contain requirements. They provide guidance to users of the International Standard and are not intended to add to or modify requirements.

The titles of the annexes are as follows:

- Annex A, Clarification of new structure, terminology and concepts

- Annex B, Other International Standards on quality management and quality management systems developed by ISO/TC 176

We recommend absorbing the content of the annexes for several reasons:

- Many nuggets of information are embedded that can potentially facilitate implementation of compliant processes

- Insight into the intent of the requirements may be acquired to enhance efficient and value-adding compliance

- Improved clarity may be acquired to better understand the content in clauses 4–10 that may seem vague

- Ideas may be obtained to assist quality professionals to "sell" process changes to the QMS of an organization that may be needed to achieve compliance with requirements

- Insight may be obtained into the rationale for the structure and form of the changes that were made that will facilitate explanation of the reasons for what appears to be a major change to the International Standard

- There are references to many other sources of supporting information that can be useful not only for achieving a minimal level of compliance but also for expanding the breadth and depth of the QMS of an organization

- Table A.1 in Annex A shows the major differences in terminology between ISO 9001:2015 and the previous edition, which may expedite consideration of any QMS changes

- Table B.1 in Annex B provides insight into the relationship between the clauses of ISO 9001:2015 and the other International Standards on quality management and QMSs

A few of the nuggets we derived from perusing the annexes that may be helpful to keep in mind when reviewing ISO 9001:2015 and contemplating process changes or additions to ensure compliance with its requirements include the following:

- There is no requirement to apply the structure and terminology used in ISO 9001:2015 to the documented information of an organization's QMS.

- The structure of clauses is intended only to provide a coherent presentation of requirements, rather than to be used as a model for documenting an organization's policies, objectives, and processes.

- There is no requirement for an organization to replace the terms it uses with the terms used in ISO 9001:2015.

- Organizations can choose to use terms that suit their operations, such as using "records," "documentation," or "protocols" rather than "documented information"; or "supplier," "partner," or "vendor" rather than "external provider."

- Clause 4.2, on interested parties, does not extend QMS requirements beyond the scope of ISO 9001:2015. As stated in the scope, ISO 9001:2015 is applicable where an organization needs to demonstrate its ability to consistently provide products and services that meet customer and applicable statutory and regulatory requirements, and aims to enhance customer satisfaction.

- There is no requirement in ISO 9001:2015 for the organization to consider interested parties where it has decided that those parties are not relevant to its QMS.

- It is for the organization to decide whether a particular requirement of a relevant interested party is relevant to its QMS.

- ISO 9001:2015 does not have a separate clause or subclause on preventive action. The concept of preventive action is expressed through the use of risk-based thinking in formulating QMS requirements.

- Although clause 6.1 specifies that the organization shall plan actions to address risks, there is no requirement for a comprehensive process for risk management or for related documented information.

- Organizations can decide whether to develop a more extensive risk management methodology than is required by ISO 9001:2015.

- Under the requirements of clause 6.1 the organization is responsible for its application of risk-based thinking and the actions it takes to address risk, including whether to retain documented information as evidence of its determination of risks.

- ISO 9001:2015 does not refer to "exclusions" in relation to the applicability of its requirements to the organization's QMS, but the organization can decide that a requirement is not applicable only if its decision will not result in failure to achieve conformity of products and services.

- The term "documented information" is used for all document requirements.

- ISO 9001:2015 addresses the need to determine and manage the knowledge maintained by the organization, to ensure that it can achieve conformity of products and services, so it has introduced requirements regarding organizational knowledge.

- All forms of externally provided products and services are addressed by ISO 9001:2015, and the organization has latitude in implementing the controls required for external provision, which can vary widely depending on the nature of the products and services.

- A matrix showing the correlation between the clauses of ISO 9001:2015 and the previous edition (ISO 9001:2008) can be found on the ISO/TC 176/SC 2 open-access website at http://www.iso.org/tc176/sc02/public.

- There is no reference to the sector-specific QMS standards that incorporate some or all ISO 9001:2015 requirements.

The important point to remember when creating new or implementing changes to processes is that you decide how to best address the needs of your organization and your customers.

Given the vagueness of some of the requirements or the absence of a requirement to have documented information for a process, there is potential for issues related to assessing conformity, especially with outside auditors (i.e., registrars). So we recommend care in ensuring that your rationale for asserting conformity is clear and able to be substantiated and that your organization is prepared to defend its position of conformity with outside auditors.

The bibliography to ISO 9001:2015 lists 26 ISO or IEC standards that organizations may find useful in their efforts to create or modify or update their QMS. We recommend review of the bibliography to determine whether information or guidance is available to enhance the content of your QMS or at least facilitate process development where needed.

In addition to the references listed in the ISO 9001:2015 bibliography, we recommend consideration and use of several of our publications from ASQ, such as:

- *How to Audit the Process-Based QMS*

- *Unlocking the Power of your Quality Management System: Keys to Business Performance Improvement*

- *ISO 9001:2008 Explained*

- *ISO 9001:2008 Explained and Expanded*

These books can provide quality professionals with additional reference material to develop and deploy efficient and effective QMS processes.

All users of ISO 9001:2015 should also become familiar with and use the information available on the ISO TC 176/SC2 public website, which is http://www.iso.org/tc176/sc02/public.

11

Auditing Implications

*No pleasure is comparable to the standing upon the vantage ground
of truth and to see the errors and wanderings and tempests in the vale
below.*

—Francis Bacon

INTRODUCTION

Auditing, by definition, is a process in which an objective and
impartial evaluation is made of all or part of a QMS's implementation against agreed-upon criteria. There are several kinds
of audits that occur related to ISO 9001, including internal
audits, supplier audits, and registration audits. This chapter
addresses the implications of internal audits and registration
audits related to ISO 9001:2015.

Internal quality audits are performed by or at the direction
of members of an organization to evaluate the effectiveness of
system implementation and whether QMS requirements are
being met. Internal audits can also be used to identify opportunities for improvement, although such activity is not an explicit
requirement of ISO 9001.

Registration audits are performed by third-party auditors
(i.e., individuals not affiliated with the organization or the processes being audited). Auditors that provide registration audit

services represent organizations called certification bodies or registrars—organizations accredited to perform audits as part of their certification or registration scheme. If your organization is seeking certification, we recommend working only with an accredited registrar and an accreditation body that is a member of the International Accreditation Forum (IAF) such as the ANSI/ASQ American National Accredited Board (ANAB).

For several reasons, the auditing of processes and QMSs that are designed and deployed to address ISO 9001:2015 requirements has the potential of a greater degree of uncertainty than prior auditing activities. This chapter explores possible issues that may arise and their implications related to both internal audits and external registration audits conducted to assess conformity to the requirements of ISO 9001:2015.

INTERNAL AUDITS

Although there are no substantive changes in the internal audit requirements from the prior editions of ISO 9001, the internal audit process of an organization may require modification due to several changes in other areas of the 2015 edition of ISO 9001.

Some of the changes in ISO 9001:2015 that could impact the internal audit process include:

- The increased amount of non-prescriptive requirements

- Vague or wishy-washy statements of requirements

- Changes in the requirements for *documented information* (i.e., procedures and records)

- An increased emphasis on applying the *process approach* to the QMS

- An updated edition of the QMPs

- Actions required to address *risks and opportunities*

- Increased emphasis on *leadership* activities
- Understanding the *organization and its context*
- Understanding the *needs and expectations of interested parties*
- Auditing *knowledge* requirements
- Auditing processes where ISO 9001:2015 does not require a process to be described in documented information (i.e., in a procedure)

Our comments and actions an organization may consider for tweaking its internal audit process in each of these areas are discussed in the following paragraphs.

The Increased Amount of Non-Prescriptive Requirements

The latest edition of ISO 9001 has removed previous requirements in several places. Examples include (1) no requirement for a management representative, (2) no specific requirement for a quality manual, and (3) requirements related to measuring equipment and calibration. Many of these changes were intended to make ISO 9001:2015 more friendly to the non-manufacturing segment of users (e.g., software and service organizations).

Organizations should consider how to train internal auditors to be able to assess what controls and documented information and objective evidence are required to meet internal requirements as well as the requirements of ISO 9001:2015. Even if there is not an explicit requirement in ISO 9001:2015, the organization may need or desire to have procedures and controls to ensure conformity of products and services to customer requirements and to facilitate consistent performance of activities. ISO 9001:2015 contains the minimum requirements for documented information, but more may, and in most cases will, be required.

Internal auditors should be trained to look for the controls and documents and records to provide objective evidence that customer and internal requirements will be met consistently. Internal audit administrators will need to consider how much time will be required to perform internal audits, which may be longer than previous internal audits. The choice of internal auditors may also need to be revisited since there could be more uncertainty in attempting to address the adequacy and conformity of processes. Auditing will become more complicated than an exercise to check off a series of boxes.

Vague or Wishy-Washy Statements of Requirements

Internal auditors and the entire organization will have to address content in ISO 9001:2015 that is expressed in terms like "relevant," "intended," "consider," "review," and "monitor." How does an internal auditor decide whether process owners have adequately "monitored" or "considered"? How far does a process owner need to go to evaluate "relevant" risks? We suggest that internal audit administrators develop work instructions to provide internal auditors guidance on how to determine conformity with requirements when requirements are vague.

Changes in the Requirements for Documented Information (i.e., Procedures and Records)

Although the requirements for documented information included in ISO 9001:2015 may appear to be diffused, the organization is still required to ensure that documented information be controlled.

An organization should, in particular, consider its approach to ensuring conformity with the requirement in ISO 9001:2015 that states that the QMS for the organization shall include

documented information determined by the organization as being necessary for the effectiveness of the quality management system

This means that the organization needs to determine what documented information is required to conform to customer, regulatory, and internal requirements and to ensure control of the availability, suitability, retrievability, and maintenance of that documented information.

ISO 9001:2015 also requires that controls be implemented for required documented information that arises from outside the organization (e.g., industry standards, customer specifications, international and national standards) and for the retention and disposal of documented information.

Internal auditors need to be trained in the nuances of documented information requirements and changes the organization makes (if any) to the terminology it uses related to procedures, records, and documented information. In some places, documented information requirements are stated in a way that, in prior editions, would have been a requirement for a documented procedure. In other places, documented information is used in a way that in prior editions would have been stated as a requirement for a record.

This area may benefit from a work instruction to provide internal auditors with the guidance they need to determine the existence of required documented information, no matter what it is called. It may also be advisable to clarify throughout the organization the terminology that is used.

We recommend careful consideration of making any changes to achieve consistency with the new nomenclature. In a number of industries the distinction between documented procedures and records is ingrained, so continued use of the historic terminology of documented procedures and records may be prudent. In other words, do not change unless your organization has a good reason to change. For most organizations, we see *no* value in rushing to embrace the new language for internal use and no requirement to do so.

The effective auditing of documented information can be of special importance for organizations operating in regulated markets (e.g., medical devices). It is also important for organizations to have adequate documented information (i.e., procedures and records) to address product liability issues.

An Increased Emphasis on Applying the Process Approach to the QMS

This edition of ISO 9001 reinforces and emphasizes the adoption of a process approach when developing, implementing, and improving the effectiveness of the QMS. It includes specific requirements that are essential to the adoption of a process approach, including the following:

- Determine the needed processes and their application

- Determine the inputs and outputs for the processes

- Identify the sequence and interaction of these processes

- Define all the activities needed to ensure effective implementation and control of processes

- Ensure the effective operation and control of these processes

- Provide the needed resources

- Ensure process owners are assigned

- Assess risks and opportunities and actions taken, if any, to address them

- Monitor, measure, and evaluate processes, as appropriate

- Control process changes

- Pay continual attention to improving the processes of the QMS as well as the overall QMS

ISO 9001:2015 also requires the organization to maintain documented information (i.e., procedures and records) to provide evidence that processes are operating under controlled conditions.

Internal auditors may require a brief or extensive refresher on the process approach to ensure that, when auditing the process, they are assessing characteristics such as process inputs, activities, and outputs and reviewing adequacy of documented information, employee competence, and improvement initiatives and the many other items in their audit plan.

In Chapters 2–9 and 12, the process approach is addressed, including the activities that internal auditors should consider to assess effectiveness. Since this aspect of the QMS is very important, we repeat here some of the kinds of questions internal auditors should consider when assessing the application of the process approach in the organization.

Typical questions related to the inputs to a process

• What are the inputs?

• Do inputs meet specified requirements? How do we know?

• Who are the suppliers of the inputs?

• How are the input requirements defined?

• How is internal supplier performance measured?

Typical questions related to the activities occurring in a process

• Who is the customer of the process?

• What does the customer want?

• Is there an understanding of what is necessary to meet (or exceed) customer requirements?

• Are individuals performing work to ensure requirements are met?

• Do workers know what to do and have the means to do it, including documentation, time, and tools?

• Are responsibilities clear?

- Are procedures, if applicable, available? Understood?

- What are the specific requirements for successful completion?

- How do individuals know that they have performed work to meet requirements?

- What is done with the data that are collected? Who analyzes the data? Is there evidence of data analysis?

- How is continual improvement addressed?

- Is appropriate objective evidence available to support claims of compliance with requirements?

Typical questions related to process outcomes

- Is the process effective in achieving required results?

- How is conformance to customer or specification requirements determined?

- Is the process continually improved?

- How is customer feedback (external or internal) solicited and used?

Typical questions related to process failures

- What happens when deviations from requirements are found? What are the processes for:

  — Correction?

  — Control of nonconforming product?

  — Disposition of nonconforming product?

  — Analysis for possible corrective action?

  — Corrective action?

  — Use of the data for preventive action, when applicable?

An Updated Edition of the QMPs

The QMPs are not part of the requirements of ISO 9001:2015, but they are foundational building blocks. All internal and external auditors need to know and understand them.

In 2015, the QMPs were updated. Several principles were tweaked, and *Process Approach* and *System Approach to Management* were combined into one principle—*Process Approach*.

Internal auditors need to review and understand the statements of the principles and the rationale behind each principle. When performing internal audits, consistency with the QMPs should be considered and any deviation noted as at least an observation, if not a nonconformity.

Actions Required to Address Risks and Opportunities

The 2008 edition of ISO 9001 makes no mention of the term "risk," whereas the 2015 edition mentions "risk" many times. In the 2008 edition of ISO 9001, "opportunities" was mentioned 2 times, and in the 2015 edition "opportunities" is mentioned 12 times.

The use of the terms "risk-based thinking" and "risks and opportunities" is new. On the other hand, such preventive thinking and action have been required by earlier versions of ISO 9001 under the term "preventive action."

In terms of specificity, the requirements related to risks and opportunities are somewhat vague, and conformance may require new processes for some organizations. Auditor training to clarify requirements and management expectations regarding how to conform with requirements related to risks and opportunities will be advisable to obtain any value from internal audit of these processes.

Increased Emphasis on Leadership Activities

The leadership content of ISO 9001:2015 is much more explicit and expanded than the content in prior editions. The intent is to broaden and deepen the requirements so that top managers become more engaged in QMS planning, operations, and improvement. ISO 9001:2015 contains a listing in clause 5 and other places of requirements for top management to address to demonstrate leadership and commitment to the QMS. Earlier versions of ISO 9001 did address a number of these concepts, but this edition drives them home.

Internal auditors will need to be trained to be more robust in the auditing of leadership activities for several reasons, not the least of which being that a lack of top management involvement in the activities of the QMS has been a chronic issue.

Understanding the Organization and Its Context

The requirements for understanding the organization and its context mean that the organization should know itself and the external organizations and factors that affect it. Achieving such understanding forms a basis for strategic planning for the organization and its QMS. Deploying processes to address this requirement may be old hat to some organizations but new to others.

An issue internal auditors will face will be how far should an organization go to address this requirement. They also may be required to obtain a foundational knowledge of strategic planning and how to assess whether adequate attention has been dedicated to considering the interests of external parties.

Internal auditing of this aspect of ISO 9001 will be an interesting challenge to even experienced auditors. It may be a good assignment for the best and brightest due to its "newness" and its strategic characteristics.

Understanding the Needs and Expectations of Interested Parties

See the earlier comments relating to auditing the organization and its context. The biggest challenge for internal auditors will be in assessing whether the processes to address the needs and expectations of interested parties go too far, don't go far enough, or go in the wrong direction.

Internal auditors will need guidance from the audit administrator regarding how to audit these processes.

Auditing Knowledge Requirements

Requirements in the area of knowledge are vague, but the processes are important.

Our recommendation is that organizations consider documenting processes for addressing knowledge requirements, if any, and look beyond the rather nominal explicit requirements to the intent of the requirements. When requirements are clarified by the organization (in documented information such as a procedure), internal auditing of conformity will be facilitated.

Auditing Processes Where ISO 9001:2015 Does Not Require a Process to Be Described in Documented Information (i.e., in a Procedure)

There are several places in ISO 9001:2015 that do not require documented information, for example, clauses 7.1.5, 7.1.6, 7.2, 8.2.1, 8.2.2, 8.2.4, 8.4, 8.5, and 8.7.

How will auditors proceed where requirements may be amorphous? Our recommendation is that internal auditors be trained to determine whether processes exist and are being consistently implemented. Some thoughts for achieving results can be seen in Figure 11.1. Internal auditors will face challenges in addressing processes without documented information, we believe, but they will be much less severe than the issues that can arise when external auditors attempt to reach decisions about conformity of such processes (see discussion of this issue in "Assessing Conformity When No Documented Information Is Required").

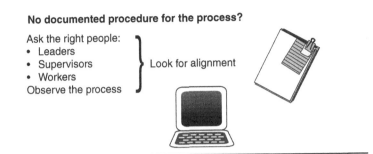

No documented procedure for the process?

Ask the right people:
- Leaders
- Supervisors
- Workers

Observe the process

Look for alignment

Figure 11.1 Auditing vague requirements.

EXTERNAL REGISTRATION AUDITS

ISO 9001:2015 is going to have some impact on how registration audits will be conducted. We believe this will be the case for several reasons, including (1) the new structure of ISO 9001:2015 (which has significant content derived from Annex SL as dictated by ISO), (2) uncertainty about what documented information is required, (3) vagueness of at least some of the requirements, (4) open-endedness of some of the requirements (e.g., determine interested parties, determine risks that need to be addressed), (5) auditors needing to "dig" to assess conformity in the absence of a procedure or a record and draw conclusions on the basis of corroborating observations and auditee statements, and (6) possibly extended discussions between auditor and auditee regarding the requirement and how conformity is being addressed.

For these reasons and for several others, we believe external audits will take longer than in the past. Audit administrators should anticipate quotes and costs for registration services to be higher than in the past due to more audit days being requested by registrars.

Some of the issues we foresee for organizations when interacting with outside auditors performing certification audits include:

- Untrained or inadequately trained registrar auditors

- Overly aggressive auditors

- External auditors who are too hard or too soft

- Arbitrary assessment by external auditors (e.g., what adequately constitutes "consider," "determine," "relevant," or "monitor")

- Assessing conformity when no documented information is required

Our recommendations for interacting with external auditors when confronting such issues are as follows.

Untrained or Inadequately Trained Registrar Auditors

The degree of auditing expertise has historically been variable between registrars and even within a registrar in spite of Exemplar Global (ANAB) and the International Accreditation Forum (IAF) efforts to establish and enforce training and competence criteria for auditors performing registration audits.

With the publication of ISO 9001:2015 we believe that this situation will be exacerbated both between registrars and within registrars, which can lead to disagreements between auditors and the organization regarding what ISO 9001:2015 says, what it means, and what constitutes conformity or nonconformity.

The best defense an organization has in interacting with registrar auditors is to be well versed in the fundamentals and principles of ISO 9001:2015 and in understanding how its QMS has been structured and deployed to meet or exceed requirements. When disagreements occur with external auditors, the prudent course of action is to discuss the rationale of your organization and the reasoning for your position and listen carefully to the position of the external auditor. It is not uncommon for external auditors to have concerns based on a lack of experience or understanding regarding how an organization has decided to implement a QMS process or the implemented process is not the way an external auditor believes a process should be employed. After both parties listen and discuss, either agreement is reached or a difference of opinion still exists. In the case of the latter, we suggest the organization not argue with an auditor. Rather, request the auditor to document both positions in the audit report for resolution by others or by the appeal process available to the organization. It is hardly ever advisable to argue with an external auditor.

An additional action an organization can employ is to remind (very gently and diplomatically) an external auditor that the auditor role is to assess conformity to a requirement, and not to assess desirability or efficiency of the processes of the organization.

Overly Aggressive Auditors

It occasionally occurs that an outside auditor is overly aggressive in performing auditor duties. We recommend that the response to such an auditor be along the lines described earlier for dealing with untrained or inadequately trained registrar auditors. Alternatively, an organization always has the option of requesting a different auditor from a registrar if there is a concern with the assigned auditor for any credible reason.

External Auditors Who Are Too Hard or Too Soft

An auditor who is too soft is as undesirable as one who is too hard. ISO 19011 (*Guidelines for auditing management systems*) outlines the personal behavior criteria for auditors. If external auditors appear to be departing from the path of objectivity and impartiality when assessing conformity to criteria, a gentle reminder of the criteria stated in clause 7.2.2 of ISO 19011 can help restore balance to the audit process.

Arbitrary Assessments by External Auditors

What is adequate evidence the organization has "considered" or "determined"? What makes something "relevant"? What constitutes adequate "monitoring"? The 2015 edition of ISO 9001 contains, in several places, criteria that may be subject to wide differences of opinion regarding what constitutes conformance. Our recommendation is that the organization clearly state the basis of conformance, and if differences of opinion cannot be resolved, then request the auditor document both positions in the audit report for resolution by others or by the appeal process available to the organization.

Assessing Conformity When No Documented Information Is Required

Although the form and words may appear different, the requirements for control of documented information (i.e., procedures and records) are similar to the requirements in ISO 9001:2008. There are, however, a few areas of difference that can lead to issues with outside auditors, such as (1) the distinction between documents and records is blurred, (2) the requirements may appear to be less restrictive in some areas, and (3) processes for which no documented procedures are required.

To avoid confusion in the organization and with outside auditors between procedures (which can be changed) and records (which can be corrected but cannot be changed), we strongly recommend organizations clearly differentiate between procedures and records. Further, we recommend that documented information (i.e., records and procedures) exist to provide objective evidence of conformity with ISO 9001 requirements as well as with internal requirements, even in places where documented information is not required. The organization is required to provide the necessary documented information (procedures and records) to ensure effectiveness of the QMS (see 7.5.1b).

To avoid conflict with external auditors when they assess conformance of processes that do not require documented information, we recommend creation of documented information (e.g., a procedure) that indicates the requirements of the organization. The documented information maintained should be in a form that suits the circumstances. There are many options ranging from documented procedures for key management system processes to simple flowcharts to work instructions. Remember, the purpose of maintaining such documented information is to enhance consistency of conformance to requirements. The right documented information will avoid uncertainty about what is required and whether processes conform to the requirements of ISO 9001:2015 and the organization.

Taking such actions will minimize conflicts with auditors and enhance internal conformity with requirements.

SECTOR IMPLICATIONS

Internal audits and external audits to sector requirements should not have any implications for your organization beyond those noted earlier for external audits to ISO 9001:2015 requirements. The organization should, as was true in the past, anticipate more intense assessment of areas that contain sector-specific requirements that are beyond the scope of the minimum QMS requirements contained in ISO 9001:2015.

SUMMARY

ISO 9001:2015 will introduce an increased level of challenge to both internal and external auditors. Your organization can minimize the level of the challenge by careful planning and deployment of processes; having a QMS that is integrated with the vision, mission, goals, and objectives of your organization; knowing and understanding the requirements of ISO 9001:2015; and being able to demonstrate how consistent conformance is being achieved. Availability of appropriate documented evidence of process requirements and conformance (i.e., records) can minimize questions related to conformity with an added potential benefit of reducing audit time.

12

Integrating the Process Approach and Systems Thinking

*Almost all quality improvement comes via simplification of design,
manufacturing . . . layout, processes, and procedures.*

—Tom Peters

BACKGROUND

The concepts of the process approach and the systems approach to management have been used by many organizations since well before the initial version of ISO 9001 was published in 1987. The systems approach to quality management can be traced back decades. Process management was a key to understanding the management systems of the 1970s. The concepts were introduced to the ISO 9000 series with the 2000 versions of ISO 9000 and 9001. The question many of us ask about ISO 9001:2015 is "Are there any new requirements?" or "What has changed?" This stuff is not new, but there is a new emphasis on the big-picture thinking of the systems approach without losing the detailed control that comes with managing the processes.

THE QUALITY MANAGEMENT PRINCIPLES

As the first QMPs were developed as a basis for ISO 9001: 2000 in the 1990s, considerable thought was given to both the notion of process management and the broader concepts related to holistic management of systems by understanding and managing process interactions. Consensus could not be reached on a single principle to describe these two notions. With the development of the current QMPs, the two ideas have been put together under the title "process approach" but with descriptive language to include the broader notion of managing the whole system. This notion of using big-picture thinking is supported by the requirements related to organizational context and risk-based thinking. Table 12.1 provides a comparison of the old and new QMP for the process approach.

Table 12.1 Principles.

ISO 9000:2005	ISO 9000:2015
Process approach: A desired result is achieved more efficiently when activities and related resources are managed as a process.	Process approach: a) Statement Consistent and predictable results are achieved more effectively and efficiently when activities are understood and managed as interrelated processes that function as a coherent system.
System approach to management: Identifying, understanding and managing interrelated processes as a system contributes to the organization's effectiveness and efficiency in achieving its objectives.	b) Rationale The QMS consists of interrelated processes. Understanding how results are produced by this system enables an organization to optimize the system and its performance.

WHAT DOES IT MEAN?

The organization's leaders need to understand how the processes, resources, and process interactions create the organization's output. In the broadest sense, the quality of that output drives market share and ultimately cash flow. Good systems with benchmarked processes are a key component of organizational success and sustainability.

To understand this concept, let's look at a simple example. Figure 12.1 represents a simple system with several processes and a number of interactions. You will notice that the number of formal interactions (i.e., the ones known well enough to be mapped) is greater than the number of processes. In this example there are five processes and eight interactions. Think about problems and nonconformities. They can occur within individual processes or at process interactions. So we not only need to address the individual processes so that they are capable of meeting output requirements, but also need to manage the interactions so that the final output of the system meets our goals.

The system illustrated in Figure 12.1 has several processes (shown as boxes) and a number of interactions (shown with arrows). The little circles in the illustration represent points in the system where key measures of performance should be tracked. It all looks to be, and conceptually is, simplicity itself.

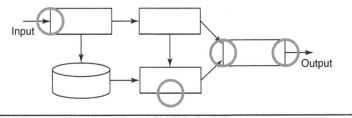

Figure 12.1 A simple system with several processes.

But a closer look at a single process within this simple system can be revealing.

But, you say, we have flowcharts of all our key processes and we understand the flow of work from one process to another. Well, let's think deeper. Figure 12.2 represents an example of a single process and its supporting elements. Notice that the process itself is straightforward but the supporting interactions can be quite complicated. The more we expect of the process, the more these supporting and interacting elements increase complexity.

As we add more activities to the process, we add internal complexity, and it becomes almost certain we will need to add some additional controls and consider additional interactions. The more of these simple details we add, the more complex the process becomes. We have increased the detail complexity of

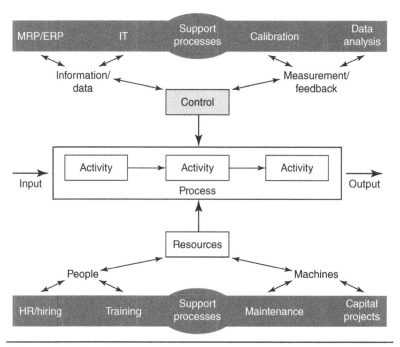

Figure 12.2   Diagram of a single process and key supporting elements.

the process and its associated system. Each interaction is clear and has a direct, understandable relationship. Managing each interface is not just possible; it is easy. But we don't have just one interaction! Operating such a process in a steady state of control, even in such a simple case, can be a daunting proposition. And, of course, the more interactions at this process level, the greater the detailed complexity.

So how do we deal with increasing complexity? We have many tools to do this. Perhaps the best is the computer, because it enables us to automate many of the interactions. At the very worst, we can just add people and "throw money at" the situation. And solving problems in such a closed process can be quite simple—using simple tools like the cause and effect diagram or even simple, detailed flowcharts. For many things, the process approach to understanding our work is almost magic. Even as the organization grows, defined linear processes make achieving great quality very doable.

But what happens when that simple linearity disappears? When the complexities of the system are no longer simple and relationships are not linear? Dynamic complexity takes over, and we can become subsumed with problems, the causes of which can be separated in time and location from the symptoms (see Figure 12.3).

In such situations, it may be impossible to know enough to draw a diagram of what the full array of interactions might

Characteristics
• Unpredictable, often nonlinear behaviors • Cause separated from effect by time and distance
Dealing with dynamic complexity
• Big-picture systems thinking and modeling—need a new paradigm • Aligned understanding and shared vision • Strategic planning

Figure 12.3 Dynamic complexity.

look like. We need to face the reality that we will require in our QMS both simple linear processes and those that are characterized by dynamic complexity.

IMPLEMENTATION TIPS

The latest edition of ISO 9001 contains new content, in the form of either new requirements or new terminology or old concepts that have been repackaged to prompt even organizations with a sophisticated QMS to pause before considering any changes. This was discussed in some detail in Chapter 1. The latest edition provides an opportunity to review where the QMS is and where it needs to be to meet the strategic objectives of your organization over the next several years. This time of review will be wasted if it does not create a simpler, more effective, and more efficient QMS. The tools of lean can be of help in some instances.

Full utilization of the process approach, including the notion of systems thinking, is a crucial part of this review. Remember that it is normal for your current system to be replete with dynamic complexity. Never be afraid to ask about the potential unintended consequences of an action—even an action you work hard to support and might not have questioned in the past.

Chapters 5–9 give advice on what to include in process management and list some of the more frequently seen interactions that may be considered to be managed. Use this information as you review and improve your system. Consider the applicability of techniques and tools such as:

- Using value stream mapping to get a good handle on the value chain and simplification opportunities

- Making a map of the overall system, its important processes, and important interactions

- Using flowcharts to study each key process

- Making serious use of failure modes and effects analysis and take real, effective action on potential problems with relatively high risk

- Measuring process capability and using control plans to ensure requirements are understood and met

- Defining and deploying a process to control the system as things change, and sticking to it

CONSIDERATIONS FOR DOCUMENTED INFORMATION TO BE MAINTAINED AND/OR RETAINED

For individual processes and the overall system, clause 4.4 requires the organization to determine what documented information is to be maintained and retained. We recommend that the process used for the system review recommended earlier be created as you start and be updated as you learn more about system development and implementation. Since a number of people will be involved, this is normally needed to drive consistency.

Information to be maintained or retained on this might include:

- A process to understand overall system interactions and efficiency (e.g., value stream mapping or other tool)

- A process for developing QMS processes

- A generic project management plan with milestones and targets that can be tweaked or customized for individual processes

- The parameters to be measured to determine the level of success achieved

WHAT IS THE END GAME?

We offer a few simple ideas as overall QMS implementation goals:

- Simplify the business model

- Simplify management systems (including, but not only, the QMS)

- Simplify the creation, internal processing, and delivery to customers of products and services

- Always remember that an overall objective is to optimize system performance and not the performance of individual processes

Organizations that achieve these goals are building a structure for survival. Those that don't may find themselves either mired in mediocrity or not sustainable.

13

Selected Sector Applications of ISO 9001:2015

Everybody gets so much information all day long that they lose their common sense.

—Gertrude Stein

Anything worth doing is worth doing slowly.

—Mae West (no relation to Jack West)

AUTOMOTIVE, AEROSPACE, AND TELECOMMUNICATIONS

There are currently over 1 million documented ISO 9001 certifications worldwide in both private and public sector organizations to increase confidence in the products and services they provide. ISO 9001 is also being used to improve control and business performance within organizations and between partners in business-to-business relations, to select suppliers, and to "screen" potential bidders on procurement contracts.

From its inception, ISO 9001 has been developed and maintained as the standard of generic QMS requirements. It is, by design, intended to apply generically across all business sectors and to all products, including hardware, software, processed materials, and services.

In many industries, meeting and exceeding the requirements of ISO 9001 has become essential to meet customer and marketplace requirements, hence the development and publication of QMS standards that are based on and derived from ISO 9001. Such standards are called sector-specific standards—they start with the ISO 9001 requirements and add, modify, or delete requirements to address the needs of the suppliers and customers in the targeted marketplace. Examples of widely recognized and mature sector-specific standards include the following:

- Aerospace—which through the International Aerospace Quality Group has published the AS9100 series of standards

- Automotive—which through the International Automotive Task Force has published the TS 16949 series of standards

- Telecommunications—which through the QuEST Forum has published the TL9000 series of standards

There are many more sector-specific standards, but for brevity will not be explored here. The degree of detail in the sector content of requirements beyond ISO 9001 varies from sector to sector. The result is that the content of sector-specific QMS standards can vary widely, from including very specific and detailed requirements to having very abstract and general requirements.

The directive to ISO Technical Committees (including TC 176) that develop MSSs that the Annex SL structure is mandatory is fostering, we believe, a migration to more abstract requirements in the basic MSSs, including ISO 9001:2015. Hence, requirement standards like ISO 9001:2015 that were intended to be easy to understand and implement may require guidance documents to understand what they really mean, and some sectors may also determine additional content is required. The question of interest to both customers and suppliers that are required or choose to comply with sector-specific standards is "What will happen when ISO 9001:2015 is published?" To provide insight into the activities and intentions of the aerospace, automotive, and telecommunications sectors, we solicited a status report from spokespersons for these sectors and are including this information as an overview of what these sectors intend to do following the publication of ISO 9001:2015. This status report was accurate as of June 2015.

AEROSPACE*

To ensure customer satisfaction, aviation, space, and defense organizations must provide and continually improve safe and reliable products and services that meet or exceed customer and applicable statutory and regulatory requirements. The globalization of the industry and the resulting diversity of regional and national requirements and expectations have complicated this objective. Organizations have the challenge of purchasing products and services from suppliers throughout the world and at all levels of the supply chain. Suppliers have the challenge of delivering products and services to multiple customers having varying quality requirements and expectations.

The aerospace industry has established the International Aerospace Quality Group (IAQG), with representatives from aviation, space, and defense companies in the Americas, Asia/Pacific, and Europe, to implement initiatives that make significant improvements in quality and reductions in cost throughout the value stream. This standard has been prepared by the IAQG.

*Information provided by Alan Daniels, The Boeing Company, for IAQG

The AS9100-series document standardizes QMS requirements to the greatest extent possible and can be used at all levels of the supply chain by organizations around the world. The intended result is improved quality, schedule, and cost performance by the reduction or elimination of organization-unique requirements and wider application of good practice. While primarily developed for the aviation, space, and defense industry, this standard can also be used in other industry sectors when a QMS with additional requirements beyond an ISO 9001 system is needed.

The AS9100-series standards include the following IAQG standards, with AS9100 being the series baseline standard:

- AS9100—*Quality management systems*—Requirements for aviation, space, and defense organizations

- AS9110—*Quality management systems*—Requirements for aviation maintenance organizations

- AS9115—*Quality management systems*—Requirements for aviation, space, and defense organizations; deliverable software

- AS9120—*Quality management systems*—Requirements for aviation, space, and defense distributors

The AS9100-series standard includes ISO 9001:2015 QMS requirements and specifies additional aviation, space, and defense industry requirements, definitions, and notes as shown in Table 13.1.

The AS9100-series standards being revised to harmonize with AS9100 and to address stakeholder inputs are:

- *AS9110.* Revisions include updating content to be more maintenance, repair, and operations (MRO) centric and to improve consistency with civil and military aviation regulatory requirements. The applicability of this standard is extended to airlines performing continuing airworthiness

Table 13.1 Highlights of key additional AS9100:2016 requirements that will be included beyond ISO 9001:2015 requirements.

Product safety	Added in carefully selected areas with consideration of current IAQG 9110 requirements to assure product safety during the entire life cycle.
Human factors	Added as a consideration in the Nonconformity and Corrective Action clause to ensure true root cause is identified and to ensure nonconformities do not recur.
Risk	Merged current IAQG 9100-series (operation) risk management requirements with the new ISO requirements on risk based thinking which permeates the entire management system.
Counterfeit parts	Introduced in carefully selected areas to establish basic requirements appropriate to the product.
Configuration management	Clause clarified and improved considerably to address stakeholder needs to specify requirements in more simplified terms.
Post-delivery support	Merged current IAQG 9100-series requirements with the new ISO requirements.

management activities in addition to existing stakeholders, such as MRO and original equipment manufacturers that provide maintenance services.

- *AS9115.* Revisions include adding content to enhance cybersecurity protections for software QMSs and recognizing cloud-based services and mobile applications.

- *AS9120.* Revisions include improved coverage of quality requirements for prevention of counterfeit and unapproved parts.

The IAQG takes the ISO certification scheme and adds additional oversight to address industry and regulatory expectations using the Industry-Controlled Other Party (ICOP) process for certification of organizations to the 9100 series of aerospace QMS (AQMS) standards.

The Other Party scheme is based on:

• The use of identical or equivalent international, sector, and national standards based on 9104

• An industry oversight system at international, sector, and national levels to ensure that scheme's requirements are fulfilled based on 9104/2

• Auditors authenticated against identical requirements based on 9104/3

• All information related to the scheme is collected in the Online Aerospace Supplier Information System (OASIS) database

IAQG publishes support materials on its website (http://www.sae.org/iaqg/) to help users understand the standards. The significant amount of information available for IAQG standards includes frequently asked questions, summary presentations, articles, specific topic discussions on new or enhanced requirements, and posted clarification. To obtain information related to the IAQG standards, see the IAQG Standards Register (http://www.sae.org/iaqg/publications/standardsregister.pdf).

AUTOMOTIVE*

The International Automotive Task Force (IATF) is an ad hoc group of automotive manufacturers and their respective trade associations, formed to provide improved quality products to automotive customers worldwide. The purposes for which the IATF was established are to:

• Develop a consensus regarding international fundamental quality system requirements, primarily for the

*Assessment of status in September 2015 by Cianfrani and West

participating companies' direct suppliers of production materials, product or service parts, or finishing services (e.g., heat-treating, painting, and plating). These requirements will also be available for other interested parties in the automotive industry.

- Develop policies and procedures for the common IATF third-party registration scheme to ensure consistency worldwide.

- Provide appropriate training to support ISO/TS 16949 requirements and the IATF registration scheme.

- Establish formal liaisons with appropriate bodies to support IATF objectives.

IATF's official position on how it will incorporate ISO 9001: 2015 into ISO/TS 16949 was not available at press time. Our belief from off-the-record conversations with IATF management is that the next edition of ISO 9001 will be incorporated in its entirety into any update of ISO/TS 16949 subsequent to the publication of the next edition of ISO 9001.

Participants in the automotive sector supply chain are advised to visit the IATF website (http://iatfglobaloversight.org) for the latest information related to the integration of ISO 9001:2015 into its suite of requirements documents for quality systems.

TELECOMMUNICATIONS*

TL 9000 is a two-part QMS standard, published by QuEST Forum, aimed at meeting the supply chain quality requirements of worldwide information and communications technology (ICT) organizations. The standard consists of TL 9000 requirements, which contain all of the auditable requirements of

---

*Information provided by Sheronda Jeffries, Cisco Systems, for QuEST Forum

ISO 9001 and additional requirements that address requirements specific to the ICT sector, such as:

- Service availability (24/7)

- Program, product, project, and service planning

- Continuity of supply

- Maintenance, including long term and end of life

- Product security

- Disaster recovery

The TL 9000 measurements define comparable performance
measurements differentiating organizations based on the following: problem report handling; on-time delivery; system
performance (outages); and quality of hardware, software, and
services. The TL 9000 Measurements Handbook is the only
industry model that provides regular performance data for
analysis against industry benchmarks and objective product or
supplier evaluations.

QuEST Forum's TL 9000 Requirements Handbook, version 6.0, will incorporate ISO 9001:2015. This was communicated in a May 27, 2014, announcement signed by the QuEST
Forum CEO, Fraser Pajak. New TL 9000 requirements supporting strategic focus areas such as security, supply chain management, corporate social responsibility, and enhancements related
to TL 9000 measurements and other QuEST Forum initiatives
are expected to be included in version 6.0.

Also, according to the International Accreditation Forum
(IAF), ISO 9001–certified organizations will be allowed three
years to transition to ISO 9001:2015. Normally for TL 9000
updates, "organizations seeking to achieve or maintain certification may continue to use the previous release of the handbook
for 12 months after the date of publication of the new release.
At that point, the old release becomes obsolete and may no
longer be used for any certification or surveillance activities."

However, for alignment with the ISO 9001:2015 three-year transition period, QuEST Forum has agreed to a two-year transition for TL 9000–certified organizations. This will also be covered in a TL 9000 Informational Alert on the rules for implementation of version 6.0 of the TL 9000 Requirements Handbook. For example, there may be a difference between the TL 9000 Requirements Handbook publication date and the effective date to ensure adequate time for TL 9000–certified organizations to complete delta training from version 5.5 to version 6.0 of the TL 9000 requirements and seamlessly upgrade to version 6.0.

A draft of version 6.0 of the TL 9000 Requirements Handbook is expected to be available for review by all QuEST Forum members by the first quarter of 2016, and version 6.0 of the TL 9000 Requirements Handbook is expected to be published by the third quarter of 2016.

To learn more about the TL 9000 Requirements Handbook update to version 6.0, visit http://www.tl9000.org or http://www.questforum.org and submit your questions to http://www.questforum.org/contact-us/.

Index

The Knowledge Center
www.asq.org/knowledge-center

Learn about quality. Apply it. Share it.

ASQ's online Knowledge Center is the place to:

- Stay on top of the latest in quality with Editor's Picks and Hot Topics.

- Search ASQ's collection of articles, books, tools, training, and more.

- Connect with ASQ staff for personalized help hunting down the knowledge you need, the networking opportunities that will keep your career and organization moving forward, and the publishing opportunities that are the best fit for you.

Use the Knowledge Center Search to quickly sort through hundreds of books, articles, and other software-related publications.

www.asq.org/knowledge-center

TRAINING CERTIFICATION CONFERENCES MEMBERSHIP **PUBLICATIONS**

Ask a Librarian

Did you know?

- The ASQ Quality Information Center contains a wealth of knowledge and information available to ASQ members and non-members

- A librarian is available to answer research requests using ASQ's ever-expanding library of relevant, credible quality resources, including journals, conference proceedings, case studies and Quality Press publications

- ASQ members receive free internal information searches and reduced rates for article purchases

- You can also contact the Quality Information Center to request permission to reuse or reprint ASQ copyrighted material, including journal articles and book excerpts

- For more information or to submit a question, visit **http://asq.org/knowledge-center/ask-a-librarian-index**

Visit www.asq.org/qic for more information.

TRAINING CERTIFICATION CONFERENCES MEMBERSHIP **PUBLICATIONS**

The Global Voice of Quality

Belong to the Quality Community!

Established in 1946, ASQ is a global community of quality experts in all fields and industries. ASQ is dedicated to the promotion and advancement of quality tools, principles, and practices in the workplace and in the community.

The Society also serves as an advocate for quality. Its members have informed and advised the U.S. Congress, government agencies, state legislatures, and other groups and individuals worldwide on quality-related topics.

Vision

By making quality a global priority, an organizational imperative, and a personal ethic, ASQ becomes the community of choice for everyone who seeks quality technology, concepts, or tools to improve themselves and their world.

ASQ is...

- More than 90,000 individuals and 700 companies in more than 100 countries

- The world's largest organization dedicated to promoting quality

- A community of professionals striving to bring quality to their work and their lives

- The administrator of the Malcolm Baldrige National Quality Award

- A supporter of quality in all sectors including manufacturing, service, healthcare, government, and education

- YOU

Visit www.asq.org for more information.

TRAINING CERTIFICATION CONFERENCES MEMBERSHIP **PUBLICATIONS**

The Global Voice of Quality®

ASQ Membership

Research shows that people who join associations experience increased job satisfaction, earn more, and are generally happier*. ASQ membership can help you achieve this while providing the tools you need to be successful in your industry and to distinguish yourself from your competition. So why wouldn't you want to be a part of ASQ?

Networking

Have the opportunity to meet, communicate, and collaborate with your peers within the quality community through conferences and local ASQ section meetings, ASQ forums or divisions, ASQ Communities of Quality discussion boards, and more.

Professional Development

Access a wide variety of professional development tools such as books, training, and certifications at a discounted price. Also, ASQ certifications and the ASQ Career Center help enhance your quality knowledge and take your career to the next level.

Solutions

Find answers to all your quality problems, big and small, with ASQ's Knowledge Center, mentoring program, various e-newsletters, *Quality Progress* magazine, and industry-specific products.

Access to Information

Learn classic and current quality principles and theories in ASQ's Quality Information Center (QIC), *ASQ Weekly* e-newsletter, and product offerings.

Advocacy Programs

ASQ helps create a better community, government, and world through initiatives that include social responsibility, Washington advocacy, and Community Good Works.

Visit www.asq.org/membership for more information on ASQ membership.

*2008, The William E. Smith Institute for Association Research

The Global Voice of Quality®